X International Geomechanics Conference (X IGC 2022)

IOP Conference Series: Materials Science and Engineering
Volume 1264

Varna, Bulgaria
18 – 23 September 2022

ISBN: 978-1-7138-6424-0
ISSN: 1757-8981

Printed from e-media with permission by:

Curran Associates, Inc.
57 Morehouse Lane
Red Hook, NY 12571

Some format issues inherent in the e-media version may also appear in this print version.

Institute of Physics
Dirac House, Temple Back
Bristol BS1 6BE UK

Phone: 44 1 17 929 7481
Fax: 44 1 17 920 0979

techtracking@iop.org

Additional copies of this publication are available from:

Curran Associates, Inc.
57 Morehouse Lane
Red Hook, NY 12571 USA
Phone: 845-758-0400
Fax: 845-758-2633
Email: curran@proceedings.com
Web: www.proceedings.com

TABLE OF CONTENTS

Preface

The X International Geomechanics Conference is jubilee edition of traditional International Geomechanics Conference that Scientific and technicalunion of mining, geology and metallurgy in Bulgaria with co-operation with other organizations and universities organized.

The International Geomechanics Conference has become a traditional scientific event and has proven its authority and importance over the years. The topics of the conference, determined by the members of the National editorial committee, are current and significant. The big challenge facing the mining industry is the rational use of mineral resources.

The X International Geomechanics Conference was held in Varna, Bulgaria, in the period 19-23 September 2022. The number of participants is 153 scientists and experts from the mining companies from Bulgaria, Germany, Japan, North Macedonia, Serbia and Ukraine. The participation was available online also for the participants that have not opportunity to arrive in Varna.

The Organizing committee expresses the gratitude to the sponsors of the conference: Golden Sponsor - Dundee Precious Metals, Silver Sponsors - Geotechmin OOD, Ellatzite Med AD, Geotrading AD, Geostroy AD and the Bronze Sponsor - Asarel Medet AD.

The organizers thank to all participants and to the authors of the papers, presented on the Conference.

IOP Conf. Series: Materials Science and Engineering 1264 (2022) 011001 doi:10.1088/1757-899X/1264/1/011001

ORGANIZING COMMITTEE

Chairman:	
Assoc. Prof. Dr. Kremena Dedelyanova	Chairperson, Scientific and Technical Union of Mining, Geology and Metallurgy
Honorary Chairmen:	
Prof. DSc Eng. Tzolo Voutov	Honorary Chairman, Scientific and Technical Union of Mining, Geology and Metallurgy
Prof. DSc Nikolay Nikolaev	Chairman, National Committee of Rock Mechanics
Deputy Chairmen:	
Prof. DSc. Eng. Nikolay Valkanov	Chairman, Bulgarian Chamber of Mining and Geology
Prof. Dr. Ivaylo Koprev	Rector, University of Mining and Geology "St. Ivan Rilski"
Prof. Dr. Slaveyko Gospodinov	Vice-rector, University of Architecture, Building and Geodetics
Dr. Eng. Vladimir Vutov	Vice-chairman, Scientific and Technical Union of Mining, Geology and Metallurgy
MSc Eng. Shteryo Shterev	Vice-chairman, Scientific and Technical Union of Mining, Geology and Metallurgy
Scientific Secretaries:	
Dr. Eng. Georgi Dachev	Ministry of Energy
MSc. Eng. Ivan Avramov	Zinkgruvan Mining AB, Lundin Mining
Dr. Julian Dimitrov	Scientific and Technical Union of Mining, Geology and Metallurgy
Dr. Krassimira Arsova	Scientific and Technical Union of Mining, Geology and Metallurgy
Members:	
MSc Eng. Dimitar Angelov	Executive Director, Kaolin JSCo.
Dr. Eng. Nevena Babunska	University of Transport "Todor Kableshkov"
MSc Eng. Rumen Bikov	Executive Director, Minproekt JSCo.
Prof. Dr. Stoyan Bratoev	Executive Director, Metropolitan JSCo.
Dr. Eng. Ivan Vutov	Chairman of the Board of Directors, Geotrading JSCo.
Dr. Eng. Vladimir Genevski	Scientific and Technical Union of Mining, Geology and Metallurgy
Prof. Dr. Iliya Garkov	Executive Director, Dundee Precious Metals Chelopech EAD, Dundee Precious Metals Krumovgrad EAD
MSc Eng. Dragomir Draganov	Executive Director, Ellatzite Med JSCo.
Prof. DSc. Dimcho Iossifov	Scientific and Technical Union of Mining, Geology and Metallurgy
MSc Eng. Ivan Kozhuharov	Executive Director, Eurotest-Control JSCo.
Dr. Eng. Krasimir Karparov	Asarel-Medet JSCo
Prof. Dr. Ivan Markov	Rector, University of Architecture, Building and Geodetics
Assoc. Prof. Dr. Ivan Mitev	Executive Director, Bulgarian Chamber of Mining and Geology
Prof. DSc Valeri Mitkov	University of Mining and Geology St. Ivan Rilski
MSc Eng. Delcho Nikolov	Executive Director, Asarel-Medet JSCo
Prof. Dr. Pavel Pavlov	University of Mining and Geology St. Ivan Rilski
MSc Eng. Dancho Todorov	Executive Director, Niproruda JSCo.
Prof. Dr. Stanislav Topalov	Chairman, Association Bulgarian Coal Mining
MSc Eng. Ilza Chinkova	Executive Director, Maritza-Iztok Mines JSCo.
MSc Eng. Yasen Chaushev	Maritza-Iztok Mines JSCo.

IOP Conf. Series: Materials Science and Engineering 1264 (2022) 011001 doi:10.1088/1757-899X/1264/1/011001

IOP Conf. Series: Materials Science and Engineering 1264 (2022) 011002 doi:10.1088/1757-899X/1264/1/011002

Peer Review Statement

All papers published in this volume have been reviewed through processes administered by the Editors. Reviews were conducted by expert referees to the professional and scientific standards expected of a proceedings journal published by IOP Publishing Publishing.

- **Type of peer review:** Single Anonymous
- **Conference submission management system:** Morressier
- **Number of submissions received:** 6
- **Number of submissions sent for review:** 6
- **Number of submissions accepted:** 5
- **Acceptance Rate (Submissions Accepted / Submissions Received × 100):** 83.3
- **Average number of reviews per paper:** 1.6
- **Total number of reviewers involved:** 7
- **Contact person for queries:**
  Name: Dr. Eng. Kremena Dedelyanova
  Email: mdgm@fnts.bg
  Affiliation: President, Scientific and technical union of mining, geology and metallurgy

Strengthening optimisation of a deep construction excavation in the city of Sofia using the observational method

Assoc. prof. Eng. Andrey Totsev, PhD

atocev@gmail.com

Abstract. This paper discusses the role of the observational method in controlling the behaviour of retaining walls and providing opportunities for savings or the prevention of construction hazards. The method is standardized in Eurocode 7 as one of the defining principles of geotechnical design and is a viable verification approach for limit states. However, in Bulgaria the rate of its application hardly exceeds 10-15% of all the implemented building projects. The advantages of the method will be demonstrated using a specific building site as a real life example.

1. Introduction

In view of the construction of a residential building in the city of Sofia, the building excavation was strengthened using a three-row Berlin-type anchored retaining wall. The depth of the excavation is 10.5 metres. The anchors used are R32 passive single bar anchors with a permissible load of 280 kN and a yield strength of 230 kN. An essential feature of bar anchors is that although they do not have a distinct anchor bond length during installation, a bond length is formed during their operation (loading), thus they are considered loaded in tension. Their static scheme is assumed without considering bending and shear. The piles are made of an IPE300 2T profile.

Photos 1 and 2. Retaining wall and anchor load cells

IOP Conf. Series: Materials Science and Engineering 1264 (2022) 012001 doi:10.1088/1757-899X/1264/1/012001

2. Engineering geological and hydrogeological conditions

The lithology of the earth's surface comprises deluvial-alluvial deposits covering Neogene sediments, which are part of the Lozenets formation. The Quaternary is made up of a culture layer – Smolnitsa soils, silty clays, and multigrain gravels. Their total thickness reaches 9.0 meters. Neogene sediments are represented by silty and silty sandy clays at a depth below 7.0 - 9.0 m from the ground. The culture layer - black-brown Smolnitsa-type clays are established in the upper part of the section. Their thickness is 1.5 m. It is designated as Layer 1. Layer 2 consists of brown to rusty clays with medium to hard-plastic gravelly inclusions and is established below the culture layer. The bottom of the layer is at a depth of 3.0-4.0 m from the ground, and the thickness of the layer is 1.5 - 2.5 m. Layer 3 is made up of multigrain gravels with a sandy filler. The gravels are established at a depth below 3.0 -4.5 m throughout the entire studied area. The bottom of the layer is at a depth of 7.0 - 9.0 m from the ground. The thickness of the layer is between 4.0 and 5.0 m. Layer 4 comprises silty-sandy and silty brown, gray brown and gray clays. It is established below 7.0 - 9.0 m from the ground. The summarized properties of separate soil varieties are shown in Table 1.

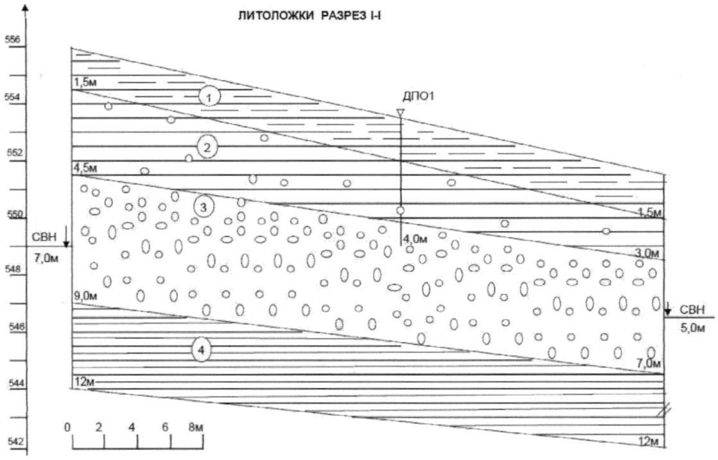

Figure 1 Geological profile

Table 1. Design values of physico-mechanical indicators - DA2 (γφ'=γc'=γγ'=1.0). – DA2 (γ_φ'=γ_c'=γ_γ'=1,0).

Layer	Volume weight	Internal friction angle	Cohesion	General deformation modulus
	γ_k [kN/m^3]	φ_k [°]	c_k [kPa]	E_0 [MPa]
Lithology type 1	17.90	16.00	14.00	4.50
Lithology type 2	18.30	11.80	19.00	12.00
Lithology type 3	21.50	31.50	3.30	31.00
Lithology type 4	19.40	19.50	22.00	15.00

3. Modelling of a strengthening structure

There are different approaches when designing the retaining walls of deep construction excavations. In general, they fall into three groups [1]: FEM with different types of nonlinear soil laws; modulus of subgrade reaction methods using linear and non-linear spring stiffness distribution (SRM) and classical methods using active and passive earth pressure. In the design of this specific excavation, a SRM was used, in which a proportionality factor is used to model the soil-structure interaction considering the correlation between stresses and deformations ($p = k \times y$).

Unlike classical methods, SRM offers a solution taking into account both soil properties and wall performance. The main issue related to the use of SRM in the design of retaining walls, which is still a subject of debate in scientific and design circles, concerns the distribution of the Winkler coefficient in depth and its exact determination depending on the obtained experimental data for the specific soil massif, as well as to the magnitude, shape and distribution of the earth pressure. Three loading schemes were considered as shown in Fig. 1 – classical loading; Terzaghi-Peck and EAB [3, 4]. The springs modelling the soil under the elevation of the bottom of the excavation were assumed to be constant and linearly increasing (Fig. 1). The anchors were modelled as yielding spring supports with a stiffness of 25,000 kN/m'.

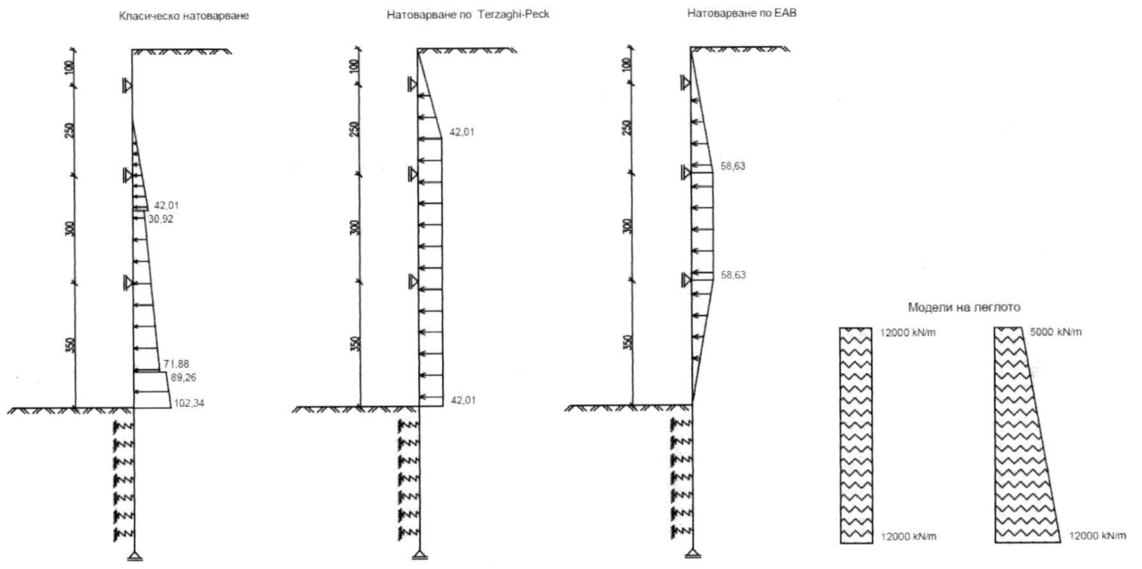

Figure 2 Retaining wall loading and support

Table 2. Calculation results

Design model		Mmax	δ1[-1.0]	δ2[-3.5]	A1	A2	A3
		[kNm]	[mm]	[mm]	[kN]	[kN]	[kN]
Classical	A=25000 и Kz=12000	75.2	5.88	11.51	2.9	64.73	189.94
Classical	A=25000 и Kz=5000-12000	84.21	4.39	13.36	2.9	61.29	203.03
Terzaghi- P.	A=25000 и Kz=12000	62.07	8.48	11.44	60.24	177.06	233.59
Terzaghi- P.	A=25000 и Kz=5000-12000	69.67	8.4	12.8	60.24	174.19	244.56
EAB	A=25000 и Kz=12000	36.95	8.2	6.06	38.03	149.59	164.25
EAB	A=25000 и Kz=5000-12000	36.25	8.16	6.17	38.03	148.64	167.93

4. Monitoring of the retaining wall

The observational method was applied by installing four measuring anchor load cells (Photo 2) on four anchors - two on the first row of anchors and two on the second row for direct measurement of the variation of the strain during the individual phases of excavation. Each load cells was developed with eight strain gauges located symmetrically along the four perpendicular axes of a steel cylinder, with four of the gauges measuring the longitudinal deformations and four compensatory gauges measuring the influence of the transverse deformations. The principle of operation of strain gauges is based on the change in the resistance of an electrical circuit when it is displaced.
Elongation or shortening of the tested structure (load cell) causes a corresponding elongation or shortening of the wire of the strain receiver (gauge), thus leading to a change in its cross-section and its stressed state. These mechanical changes cause a change in electrical resistance. The version described with four active tensors is a complete Wheatstone bridge. Additionally, by means of a system of geodetic benchmarks, the movement of the retaining wall was monitored for the individual phases of implementation. The obtained results are summarized in Table 3.

Table 3. Monitoring results

Excavation phase	δ1[-1.0]	δ2[-3.5]	A1	A2
	[mm]	[mm]	[kN]	[kN]
Excavation to elevation - 1.00 m	2	2	0	-
Excavation to elevation - 3.50 m	12	11	8	0
Excavation to elevation - 6.50 m	17	14	18.3	99.1
Excavation to elevation - 10.00 m	76	66	23.2	141.2

5. Horizontal displacement of the retaining wall

In the specialized literature, there are numerous observations of the horizontal displacements of retaining walls, on the basis of which some authors propose empirical methods for determining the expected displacement.

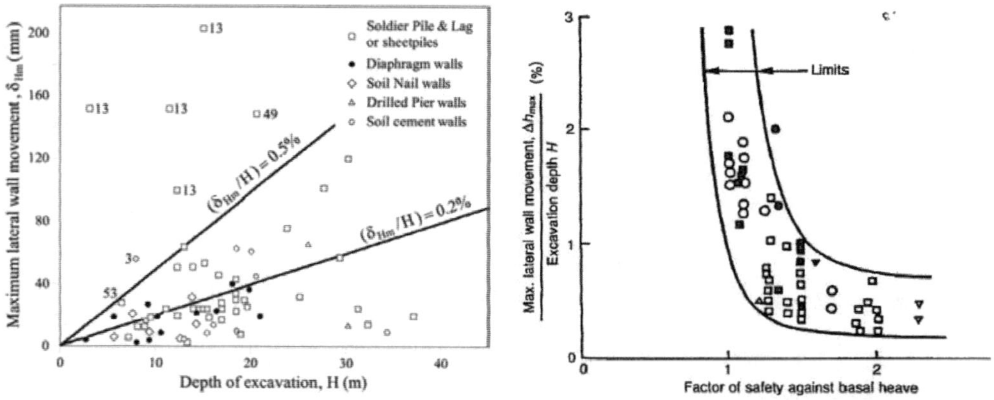

Figure 3. Determination of horizontal displacements by the empirical methods of Clough & O'Rourke for hard clays and sands (1990) and Mana & Clough for soft clays (1981) [2].

In these methods, the displacement depends on the height of the wall, the number of anchor rows and the type of soil. Figure 3 shows two such methods for different soils - Clough & O'Rourke for hard clays and sands (1990) and Mana & Clough for soft clays (1981) [2]. According to the proposed dependencies, for hard clays and sands and an excavation depth of 10.00 m, a displacement between 5 - 30 mm is expected. With weak clays and an excavation depth of 10.00 m, a displacement between 50 -100 mm is expected with a safety factor greater than 1.5.

6. Analysis of results

The observations made during the individual stages of the strengthening of the building excavation showed that the design calculation results differ from those of the monitoring results. Regarding the forces acting in the anchors, the force obtained from the anchor load cells in the first anchor row were 3 to 8 times smaller than those calculated by the different design methods. In the second anchor row, the forces were approximately twice larger than the strains obtained by the classical method and close to those obtained by the Terzaghi – Peck method and by up to 20% smaller than the results calculated by EAB. The obtained results confirm the opinion of leading specialists that the classical distribution of the load diagram does not give reliable results in multi-row anchoring. As a result of the smaller forces in the anchors obtained from the monitoring when compared with the design results, it was decided to reduce the number of anchors on the third anchor row by half by skipping every second anchor.

Regarding displacements, the results obtained from the monitoring were not close to those obtained by the design values, regardless of the chosen method. To a large extent, they were a function of the assumed stiffness of the anchors. The resulting displacement values were due to the overall stress relaxation of the massif. It is noteworthy that the results were within the limits for soft clays specified by Mana & Clough.

7. Conclusion

The presented real life construction example shows the importance and advantages of using a monitoring system. Generally, the reason not to conduct monitoring is the reluctance of most investors to direct funds for this purpose. The argument that such expenses are unnecessary is easily refutable since the monitoring cost for a specific building site amounts to approximately 3.5% of the respective reinforcement costs, and the actual savings made by reducing the number of anchors in the third anchor row amounted to 9.5% of the total reinforcement costs.

The more vital argument in support of monitoring is that it can ensure accident prevention by managing strains and displacements in a timely manner.

References:

[1] Почвени анкери, А. Тоцев, 2019
[2] Ground movement control deep excavation, Antonio Gens, NZGS course, 2010
[3] Recommendations on excavation (EAB), 2008
[4] Terzaghi K., Peck R., Soil mechanics in engineering practice New York, London, Sidney, 1967

IOP Conf. Series: Materials Science and Engineering 1264 (2022) 012002 doi:10.1088/1757-899X/1264/1/012002

BTEX exposure in indoor air in schools in Bulgaria

Ekaterina Serafimova[1], Vilma Petkova[2*]

[1]University of Chemical Technology and Metallurgy, 8, Kliment Ohridski, Blvd., 1756 Sofia, Bulgaria
[2]Institute of Mineralogy and Crystallography, Bulgarian Academy of Sciences, Acad. G. Bonchev Str., bldg.107, 1113 Sofia, Bulgaria

*E-mail address vilmapetkova@gmail.com

Abstract. Exposure to indoor air pollutants is one of the primary environmental health stressors, the most important categories of chemicals that occur in the indoor air are BTEX (benzene, toluene, ethylbenzene and xylenes). All of them are used in many commercial building products and materials use indoor. Each of the chemicals in the mixture of concern is volatile, well absorbed, extensively metabolized, and does not persist in the body for long periods of time. All of the BTEX chemicals can produce neurological impairment. The aim of this work is on research and analysis of indoor air quality. seasonal fluctuations and to assess the impact of chemical pollutants in a closed environment in southwest Bulgaria.

1. Introduction

1. A review of the literature showed that indoor air quality (IAQ) has been of scientific interest for more than three decades [1-5]. The World Health Organisation (WHO) pointed out that environmental quality is a major factor affecting children health. Many functions and organs of children are not yet well-developed and mature. It is estimated that, for example, approx. 16% of diseases and deaths registered in case of children could be caused by an inadequate quality of environmental factors [4]. It is well documented that the deterioration of IAQ is due to the appearance of a large number of chemicals or classes of chemicals in the air indoors. Among the most important categories of chemicals found in indoor air are volatile organic compounds. According to EU regulations, VOCs are classified as organic compounds with a boiling point between 50 and 260 °C. Most often these are the so-called BTEX (benzene, toluene, ethylbenzene and xylene). Main effects on human health are like primary irritation of the mucous membranes of the eyes, the upper respiratory tract, and the skin [1, 5]. Indoor air sources of carbonyls in offices buildings include building materials, plywood, adhesives, laminate floorings, paints and solvents [1-5]. In addition, it was proved that new materials and new buildings tend to emit more pollutants in air than old ones [5]. The reasons for the established spread of these chemical pollutants in the interior on the one hand is their volatile nature, and on the other - their increasing use in the manufacture of goods, products, articles and other household materials. The diversity is very large and with an increasing participation of products of human activity. This applies both to the introduction of a large number of chemical products in the raw material base, and their diverse practical application under different conditions in the premises used. The duration of use of individual products, materials and products in residential and public buildings also varies widely and this creates opportunities for additional transformations and possible formation of intermediates in the decomposition of higher molecular weight products in the process of their use. To the traditionally widely used building materials, paints, varnishes, waxes, solvents, detergents and detergents, a wide range of new composite mixtures for various purposes has been added, applied both during construction and in the furnishing

and use of the premises. In this way, potential emission generators are no longer only the coatings, but also the available furniture equipment. All this requires care to provide more and more preliminary information about the dangers of generating chemical pollutants at the stage of their selection in the various stages of construction, furnishing and use of premises, because it determines the risk to human health.

The mode of impact on humans of the generated pollutants of course depends mainly on the exposure and use of materials and products in the home, but it of course depends on the number of pollutants in the room, as well as the specific sensitivity and resistance of the inhabitants. It is natural to expect that in children and pregnant women the sensitivity is the highest, but depending on a complex of other factors, residents of other ages can be very sensitive, especially if they have any other diseases or disabilities. Other complications related to the health consequences relate to the combined exposure to several chemical pollutants, which complicates the scheme of their impact. Higher levels of BTEX are usually associated with neurological effects, most commonly associated with cancer.

The present work focuses on research and analysis of indoor air quality. To study the influence of the concentration of BTEX in indoor air, taking into account seasonal fluctuations and to assess the impact of chemical pollutants in a closed environment

2. Materials and methods

The selected pollutants were measured inside schools in the area during the heating and cooling season. The concentrations of BTEX (benzene, toluene, ethylbenzene and xylenes), were examined using a passive sampling method (Radiello-type samplers-Fig.1.).

The sampling for BTEX was conducted using passive samplers incorporated with activated charcoal absorption element. Passive samplers offer a cost–effective method for screening concentrations over a large spatial distribution. Sampling using Radiello monitors begins with a quick assembly of the support plate. The adsorbent cartridges used to collect samples are housed in a sealed glass tube that is used to store the cartridge before and after sampling. Prior to sampling, the adsorbent cartridge needs to be transferred to the appropriate diffusive body, which is then screwed onto the triangular support plate horizontally for stationary sampling, or vertically (with adapter) for personal sampling. The overall design of the Radiello sampler allows users to easily transfer the adsorbent cartridges from the diffusive body without touching the adsorbent itself. Protective outdoor shelters are recommended for environmental/ ambient air sampling. Uptake rates are dependent on temperature; therefore, concentration values obtained during sampling will be more accurate if precise temperature values are recorded during sampling. For all 7 days period of sampling was recorded a temperature in every 15 min. One sampling point per classroom was designated for indoor measurements. Samples were collected in classrooms where the children spent most of their time. The passive samplers were placed at a height of 1.5 to 2 m in the classrooms. Exposed BTEX samples were analysed using the "PerkinElmer" model of GC-FID method (Fig. 2) and standards ISO 16200-2:2002 [6] and EN ISO 16017 [7].

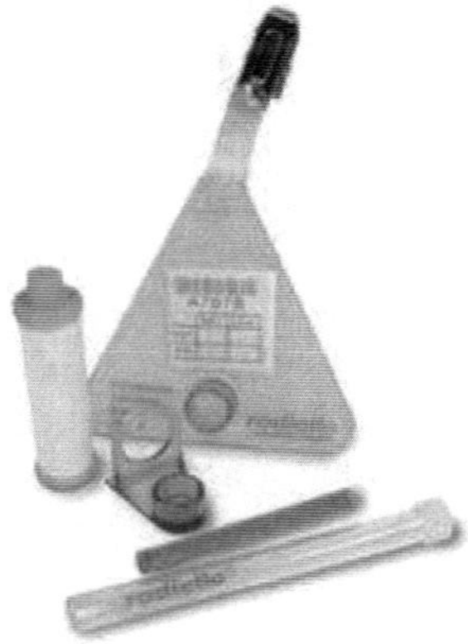

Figure 1. Radiello samplers

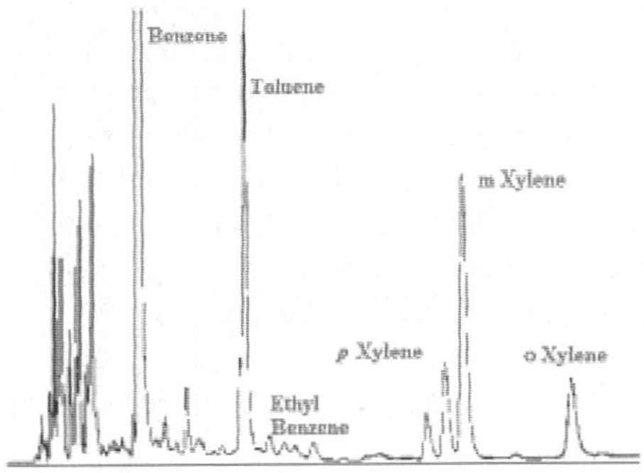

Figure 2. Result of GC-FID method

IOP Conf. Series: Materials Science and Engineering 1264 (2022) 012002 doi:10.1088/1757-899X/1264/1/012002

2.1. Sample preparation and gas chromatographic analysis

After exposure, the VOCs Caught from the cartridge with activated charcoal volatile organic compounds are extracted with carbon disulfide. Aliquots of this extract is injected into a gas chromatograph equipped with a capillary column and flame ionization detector. Quantification is done by absolute calibration.

Conditions for gas chromatography determination of benzene, toluene, xylene, ethylbenzene are as follows:

- Gas chromatograph: "PerkinElmer" model 8310 (USA) with FID detector
- Injection port: 250 °C
- Column: DB-5MS, 30m, I.D.-0,25mm, film-0.25μm
- Detector FID: 300 °C
- Column temperature: 32°C / 3 min, 8 °C min up to 100 °C.

3. Results and discussion

Indoor air quality is part of the indoor environment, which also includes physical factors and psychological aspects such as lighting, acoustics, thermal comfort and the affected age group, which is an important characteristic due to the accumulation of pollutants. As expected, depending on the age and the respective average weight of the occupants and their time spent in the premises, the respective values of the studied pollutants also change. Due to the specifics of the studied sites, information is given in Table 1 on the age of the target groups.

Table 1. Age of the target people in the studied objects

Area	Location	Object	Ages, years
Kardzhali	Kardzhali	School “Hristo Botev”	15 and up 15
	Kardzhali	School “Pencho Slaveykov”	9-14 15 and up 15
	Shiroko pole	Primary School “Ivan Vazov”	9-14
Smolyan	Zlatograd	Children. nursery “Mir”	1-2
Haskovo	Haskovo	School “I. Petkanchin”	15 and up 15
	Haskovo	Kindergarden “Zvunche”	1-2 3-8
	Karamantsi	School “Hristo Botev”	9-14 15 and up 15

Measurements of volatile organic compounds (VOCs) were carried out in a field study in in the period of January –August 2015. Seven reconstructed and repaired schools, kindergarden were chosen as sites for passive sampling of VOCs in different parts of the Municipality, taking into consideration the traffic volume profile across the city and the suburbs.

The levels of indoor air pollutants measured in the investigated classrooms in the selected schools in area are presented in table 1. The same environmental monitoring methodology was used in all the classrooms. According to the project was used the same equipment during the january-march 2015 and july-august 2015 environmental monitoring to discover some seasonal dependency

In practice, indoor exposure levels are assessed on the basis of existing guidelines and recommendations. Each EU member state sets limit values for workplace environments, but only some member states have guideline values for public places, and limit values for private spaces are very rare. Total limit value for benzene and xylene is 5 μg m^{-3}, for toluene and ethylbenzene is 3 μg m^{-3} regardless of the season.

Table 2. Seasonal concentration of BTEX in school object in Bulgaria

Objects	**Benzen, µg m^{-3}**		**Toluene, µg m^{-3}**		**Xylene, µg m^{-3}**		**Ethylbenzen, µg m^{-3}**	
	summer	**winter**	**summer**	**winter**	**summer**	**winter**	**summer**	**winter**
Kardzhali - School "Hristo Botev"	1.33	3.35	<0.18 *OMM	<0.18 *OMM	<0.24 *OMM	<0.24 *OMM	<0.21 *OMM	<0.21 *OMM
Kardzhali - School "Pencho Slaveykov"	1.02	3.42	<0.18 *OMM	<0.18 *OMM	<0.24 *OMM	<0.24 *OMM	<0.21 *OMM	<0.21 *OMM
Shiroko Pole - Primary School "Ivan Vazov"	4.84	6.97	0.73	2.90	<0.24 *OMM	<0.24 *OMM	<0.21 *OMM	<0.21 *OMM
Zlatograd - children. nursery "Mir"	3.12	1.50	1.41	3.01	2.03	5.78	7.30	8.30
Haskovo - School "I. Petkanchin"	2.66	4.01	<0.18 *OMM	<0.18 *OMM	<0.24 *OMM	<0.24 *OMM	<0.21 *OMM	<0.21 *OMM
Haskovo - kindergarden "Zvunche"	3.04	6.89	<0.10 *OMM	<0.18 *OMM	<0.24 *OMM	<0.24 *OMM	<0.21 *OMM	<0.21 *OMM
Karamantsi - school "Hristo Botev"	1.31	3.41	<0.10 *OMM	<0.18 *OMM	<0.24 *OMM	<0.24 *OMM	<0.21 *OMM	<0.21 *OMM

The data show that benzene is found again more often in the air of the studied objects from the southeast. The values during the cold period of the year are again higher. Most values are below the method error.

During the renovation in the schools there is a replacement of wooden joinery with plastic, laminate and linoleum are laid and latex paints are used everywhere for the coatings. It is logical to assume that the emitted volatile organic pollutants are a consequence primarily of the use of: - thermal insulation boards and composite adhesives; the laminate; linoleum; laminates; PVC - joinery; latex coatings, etc. However, this does not explain the domesticated topographic distribution of these pollutants, as these materials were used in the renovation of all objects.

One possible explanation is that the objects have been recently renovated, the migration processes of the materials used have been shorter. However, there are European standards in which the exposure time must be even shorter.

The increased level of benzene in winter is explained by the insufficient aeration of the premises with outside air. PVC windows permanently interrupt the natural aeration indoors, air conditioners, which are said to exist in some sites do not bring fresh air, and on cold days ventilation is limited. The increased concentration of xylene and ethylbenzene in Mir kindergarten in Zlatograd is due to the fact that the only one that after repair was laid for linoleum flooring, which obviously does not meet the necessary requirements and continues to emit VOC after more than a year after completion of repair works.

Assessment of the possibilities for secondary decomposition, which may lead to the formation of secondary emissions of pollutants into the environment have a different physical-chemical basis for volatile organic pollutants in the premises. The very nature of building materials is also different and even determines the possibilities for generating one or another type of pollutants in the environment.

All this requires special attention. The appearance of organic pollutants in the premises is mainly related to the presence of organic components in the materials used.

4. Conclusions

When renovating schools, where many different products are imported, it further complicates the distribution of emitted organic pollutants in the premises. Therefore, in one school it is possible to establish different levels of emissions of volatile organic pollutants in different rooms. This is determined by the presence of significantly different mixtures with different contents of organic chemical components. Another feature of composite materials is their gradual reduction over time, due to their transition to the gas phase and spread in the air.

The possible adverse effects of building materials on people can be significantly different depending on the type and capacity of emissions emitted by them during the entire period of use of buildings.

At the same time, it should be assumed that the negative impact is determined not only by certain levels of emissions of organic pollutants, but also their number. Increasing the number of chemical pollutants usually results in a certain increasing effect, in which the negative effects increase.

It is necessary to apply higher control when laying different types of pavements and carrying out repair works in schools and kindergartens to prevent cumulative effects of VOCs in indoor air, especially in the younger generation, which are the most vulnerable group. It is necessary to introduce additional more restrictive maximum permissible concentrations, which should be measured over a certain period of time on a seasonal basis in schools and kindergartens.

5. References

[1] WHO guidelines for indoor air quality: selected pollutants 2010 ISBN 978 92 890 0213 4, World Health Organization.
[2] Sarigiannis D A, Karakitsios S P, Gotti A, Liakos I L and Katsoyiannis A 2009 *Sensors* **9** 731–755.
[3] Sarigiannis D A, Karakitsios S P, Gotti A, Liakos I L and Katsoyiannis A 2011 *Environ. Int.* **37** 743–765.
[4] Huang L S 2017 *Journal of Environmental Protection and Ecology* **18**, 1319–1327
[5] Diodiu R, Bucur E, Galaon T, Pascu L F 2015 *Journal of Environmental Protection and Ecology* **16**, 832–838
[6] ISO 16200-2:2000 Workplace air quality – Sampling and analysis of volatile organic compounds by solvent desorption/gas chromatography – Part 2: Diffusive sampling method
[7] ISO 16017-2:2003 Indoor, ambient and workplace air – Sampling and analysis of volatile organic compounds by sorbent tube/thermal desorption/capillary gas chromatography – Part 2: Diffusive sampling

Acknowledgements. This work was supported by the Operational Program "Science and Education for Intelligent Growth", co-financed by the European Union through the European Structural and Investment Funds under grant BG05M2OP001-1.001-0008 of National Centre for Mechatronics and Clean Technology (V.P.).

IOP Conf. Series: Materials Science and Engineering 1264 (2022) 012003 doi:10.1088/1757-899X/1264/1/012003

Phase and structural investigations of the content of natural radionuclides in rock and inorganic building materials

Ekaterina Serafimova[1], Vilma Petkova[2*], Tiit Kaljuvee[3]

[1]University of Chemical Technology and Metallurgy, 8, Kliment Ohridski, Blvd., 1756 Sofia, Bulgaria
[2]Institute of Mineralogy and Crystallography, Bulgarian Academy of Sciences, Acad. G. Bonchev Str., bldg.107, 1113Sofia, Bulgaria
[3]Tallinn University of Technology, Ehitajate tee 5, 19086, Tallinn, Estonia

*E-mail address vilmapetkova@gmail.com

Abstract. The population living in areas rich in granites, minerals and soils with high concentrations of natural radionuclides is exposed to higher external radiation. Significant part of a person's time resides in buildings where he lives and works. The walls, floors and roofs of the buildings shield (weaken) the external gamma background. The aim of the work is to determine the content of natural radionuclides in rock and inorganic materials used as building materials of samples taken from different buildings in Southern Bulgaria by analysing the samples and measuring their natural radioactivity, then to compare the results with the maximum concentrations of activity in common building materials. also used in the EU.
The data obtained from the analyzes do not give grounds to expect the generation of radionuclides from natural and conventional inorganic building materials used in the studied areas.

1. Introduction

The population living in areas rich in granites, minerals and soils with high concentrations of natural radionuclides is exposed to higher external radiation. When assessing the exposure of the population to natural radionuclides, the fact must be taken into account that a significant part of a person's time resides in buildings where he lives and works. The walls, floors and roofs of the buildings shield (weaken) the external gamma background. On the other hand, the building materials (bricks, concrete, gypsum), from which the buildings and dwellings are built, contain natural radionuclides with various concentrations. When estimating the average annual effective dose for the population of the Earth, the radiation from natural radionuclides, which is obtained outdoors and indoors, is taken into account separately.

Concentrations of indoor air pollutants are determined by the ability of outdoor source pollutants to infiltrate indoors and emissions from indoor sources such as building materials, furniture, that have a natural origin [1]. The variability of new building materials and their practical application made the different levels of contamination and different indoor microenvironments can be very wide.

The content of various radioactive elements in building materials and the characterization of their crystal structures give a first idea of the possible risks to human health in their application in the construction of public, family and industrial buildings. [2, 3]. This is the reason why more and more researchers are focused on the characterization of raw materials and composite products and materials which is used in practice.

IOP Conf. Series: Materials Science and Engineering 1264 (2022) 012003 doi:10.1088/1757-899X/1264/1/012003

The aim of the work is to determine the phase and structural properties and the content of natural radionuclides of rock and building materials of samples taken from various objects in Southern Bulgaria, measuring their phase and structural properties by Powder X-Ray analysis (PXRD), wavelength dispersive X-ray fluorescence (WD-XRF) and natural radioactivity by gamma spectroscopy and determining the specific activities of ^{40}K, ^{226}Ra and ^{232}Th. The results obtained to compare with the maximum activity concentrations in total construction materials used in the EU.

2. Materials and methods

2.1. Samples preparation

For the purposes of the study, based on the specific characteristics of the geological structures in the country, objects were selected - administrative, public and private buildings, from which sampling of materials used for construction purposes and selected the exact locations for the relevant studies and determinations. They belong to southeast direction in Bulgaria. The southeastern direction includes sites from the regions in and around cities of Haskovo, Kardzhali, Zlatograd. Samples of rock materials used for construction purposes were selected from the same areas. The samples of rock and building materials are dried at 105°C until a constant weight. The samples are grinding in a ball mill to a fraction size below 3 mm. They are homogenized and reduced in an amount of about 100 grams and to the size of the fraction - 0.063 mm (Figure 1).

Figure 1. Samples of rock materials

2.2. *Methods*

The powder X-ray diffraction (PXRD) measurements were made by D2 Phaser BrukerAXS, CuKα radiation (λ = 0.15418 nm) (operating at 30 kV, 10 mA) from 3 to 70 °2θ with a step of 0.05°, 1 s/step (ground sample weight – 1.0±0.1 mg and particle size below 0.075 mm). The PDF (Powder Diffraction File, ICDD, 2001) database was used for determining the phases and minerals in the samples [4].

Samples of rock and building materials for measurement with WD-XRF are prepared in advance in the form of tablets. The content of main chemical elements in the oxides form was performed using a WD-XRF Supermini 200 (50 kV, 200 W Pd-anode, Rigaku, Tokyo, Japan) in helium atmosphere [5]. The Supermini-200 WD-XRF system has high powder (200 W, 50 kV) Pd X-ray tube, resulting in lower detection limits and shorter measurement time. Two different X-ray detectors, a gas flow proportional counter for light elements and a scintillation counter for heavy elements are used in the present study. Depending on the wavelength range, two analyzing crystals: LIF 200 (for Ti-u), PET (for Al-Ti) and RX25 (for F-Mg) were used in the investigate studies. The quantification of each element in studied coins was performed using the fundamental parameter (FP) method. Rigaku supermini 200 has a powerful FP algorithm called as Rigaku Profile Fitting-Spectra Quant X (RPF-SQX) which allows the users to achieve standardless analysis whereby all parameters are based upon theoretical equations, the fundamental parameter database and precise modeling of the detector, X-ray tube, and instrumental geometry [6].

The analysis of the content of natural radionuclides in samples of rock materials used for construction purposes and building materials from different buildings (determination of specific activity index) was made by gamma-spectrometric analysis [7] according to international standards (IEC 61452:1995) Nuclear instrumentation – Measurement of gamma-ray emission rates of radionuclides – Calibration and use of germanium spectrometers. International Electrotechnical Commission, Geneva, 1995 [8].

The gamma spectrometric system consists of an HPGe (ultrapure germanium) coaxial detector and a digital spectrometer. The detector model is GMX45P4-83-CW-SMN, №52-TN 4225A (with HPGe crystal with volume 226 cm^3 and 0.9 mm thin fiberglass window) is relatively efficient $\varepsilon r=45\%$ and energy resolution 1.33 MeV FWHM=2.1 keV.

The system is calibrated with a reference volumetric source with traceability to an international standard (calibration certificate issued by the Bulgarian Institute of Metrology, DG "National Center of Metrology").

3. Results and discussion

3.1. PXRD

Figure 2 shows the PXRD results of the samples investigation. The detected phases were calcite ($CaCO_3$) PDF#06-6528, dolomite ($CaMg(CO_3)_2$) PDF #36-0426, quartz (SiO_2) PDF #46-1045, K-feldspar ($K_5Na_5AlSi_3O_8$) PDF #84-0710, Mica PDF #88-0791, Chlorite ($Mg_6Si_4O_{10}(OH)_8$) PDF #73-2376, Plagioclase ($CaAl_2Si_2O_8$) PDF #41-1486, Portlandite ($Ca(OH)_2$) PDF#44-1481, Gypsum ($CaSO_4.2H_2O$) PDF #21-0816, Montmorillonite ($KAl_4(Si,Al)_8O_{10}(OH)_4.4H_2O$) PDF #35-0652.

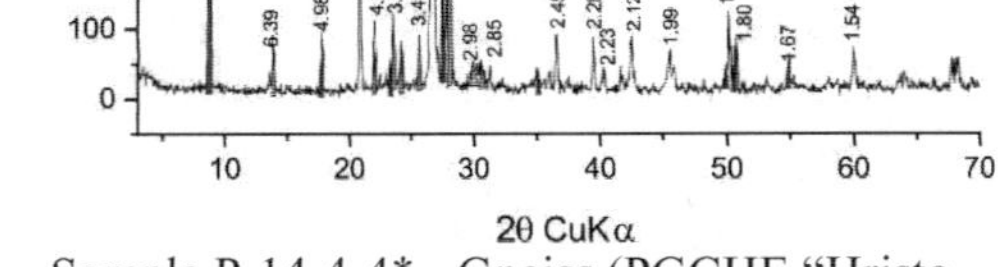

Sample P-14-4-4* - Gneiss (PGCHE "Hristo Botev"), Kardzhali city

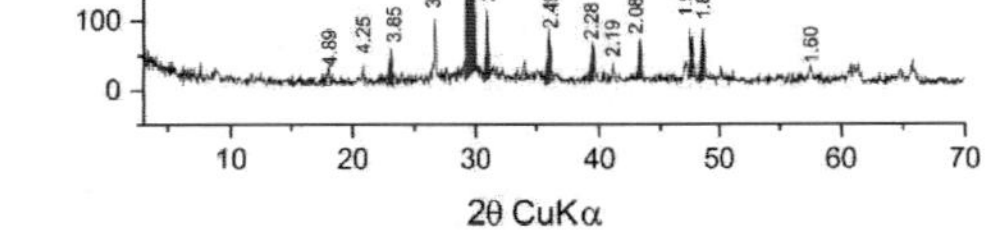

Sample P-14-2-2 - Old faience (Student Dormitories complex), Kardzhali city, Kardzhali district

IOP Conf. Series: Materials Science and Engineering 1264 (2022) 012003 doi:10.1088/1757-899X/1264/1/012003

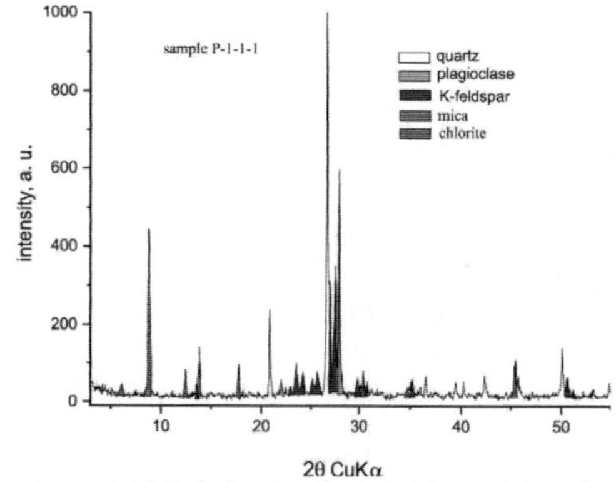

Sample P-1-1-1 - Granitoid Deformed (hotel "Ench's inns"), Zlatograd city, Smolyan region

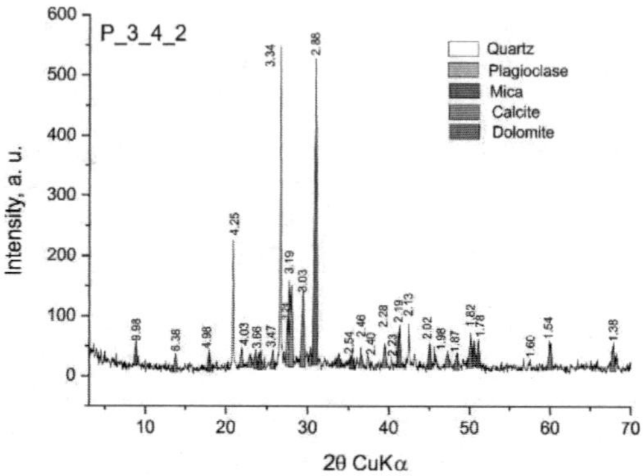

Sample P-3-4-2 - Old concrete (Mathematical High School "B. Petkanchin"), Haskovo

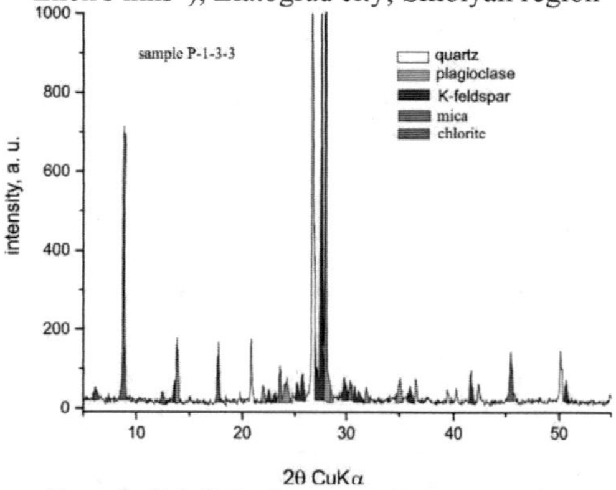

Sample P-1-3-3 - Leucocratic metagranite flooring (hospital), Zlatograd

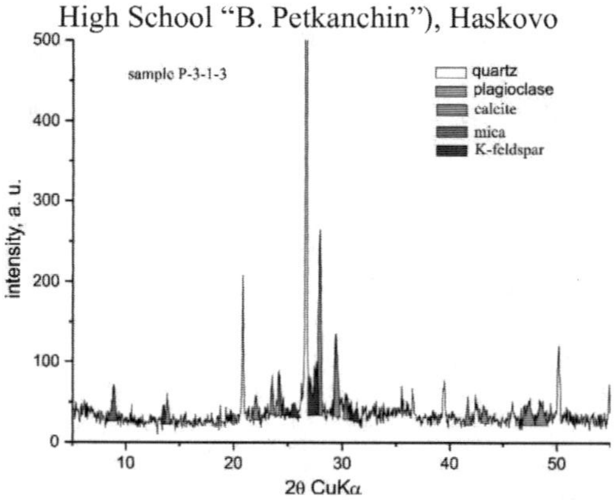

Sample P-3-1-3 - Old concrete (Sports Hall "Spartak"- chimney body), Haskovo city

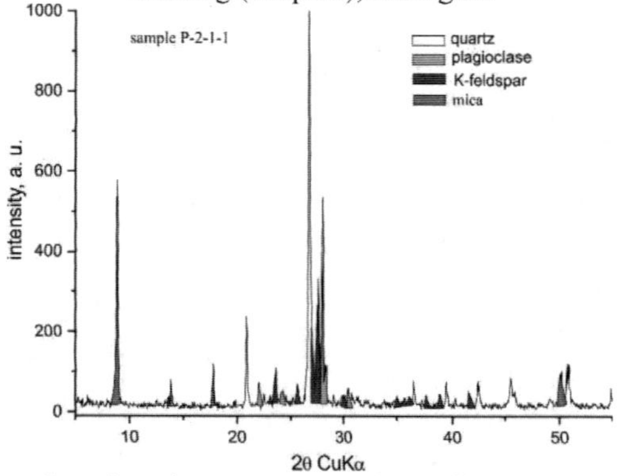

Sample P-2-1-1 - Metagranite wall cladding (rooms for rent, "Tavern at Kolcho"), Mineralni Bani village, Haskovo region

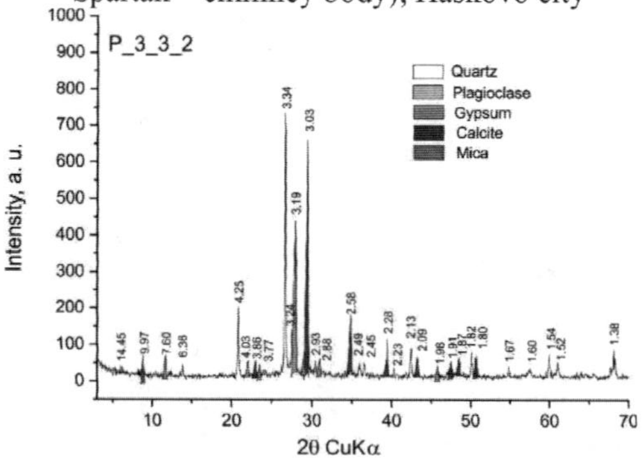

Sample P-3-3-2 - Old mortar (kindergarten "Bell"), Haskovo

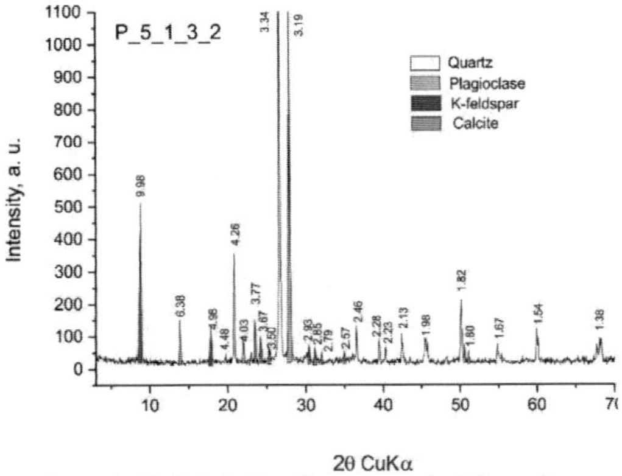

Sample P-5-1-3-2 – Granitograis (House), Vinevo

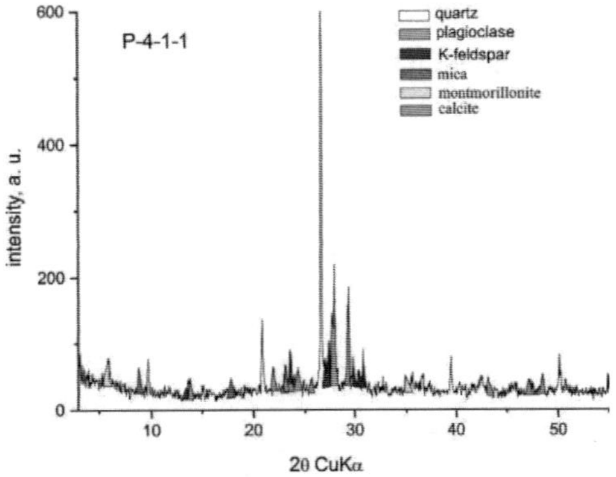

Sample P-4-1-1 – Old mortar ("Hristo Botev" High School) Karamantsi village

Figure 2. PXRD pattern of samples of rock (left column) and building materials (right column) (*Own numbering for all samples)

3.2. WD-XRF

The results obtained from the WD-XRF analyses are presented in Tables 1 and 2.

Table 1. XRF data of content of main oxides for rock building materials

No.	Component	Result, mas%				
		P-14-4-4	P-1-1-1	P-1-3-3	P-2-1-1	P-5-1-3-2
1	Na_2O	3.49	3.37	2.86	3.42	3.68
2	MgO	0.322	0.381	1.35	0.452	0.748
3	Al_2O_3	15.9	15.5	16.2	15.6	13.1
4	SiO_2	71.4	71.3	68.1	70.6	76.4
5	P_2O_5	0.374	0.337	0.150	0.340	0.107
6	SO_3	0.0276	0.0359	0.0274	0.0671	0.0088
7	K_2O	6.04	6.30	4.24	6.02	2.63
8	CaO	0.993	0.990	2.32	1.84	0.393
9	TiO_2	0.122	0.163	0.654	0.189	0.217
10	MnO	0.0363	0.0445	0.0752	0.0440	0.0209
11	Fe_2O_3	1.19	1.47	3.91	1.42	2.58
12	ZnO	0.0062	0.0063	0.0083	0.0055	0.0031
13	Ga_2O_3	0.0254	0.0038	0.0145	0.0253	0.0033
14	Rb_2O	0.0116	0.0311	0.0404	0.0121	0.0040
15	SrO	0.0069	0.0135	0.0001	0.0007	0.0030
16	Y_2O_3	0.0031	0.0000	-	0.0095	0.0092
17	ZrO_2	0.0123	0.0082	-	0.0027	0.0047
18	BaO	-	0.0761	-	-	0.0056
19	PbO	-	0.0060	-	-	0.0217

Table 2. XRF data of content of main oxides for inorganic building materials

No	Chemical Component	Result, mas%					
		P-14-2-2	P-3-4-2	P-3-1-3	P-3-2-5	P-3-3-2	P-4-1-1
1	Na_2O	3.49	3.42	2.86	1.35	0.99	3.05
2	MgO	0.322	0.452	1.35	0.803	3.13	0.428
3	Al_2O_3	15.9	15.6	16.2	6.87	16.4	12.3
4	SiO_2	71.4	70.6	68.1	38.0	54.4	64.7
5	P_2O_5	0.374	0.340	0.150	0.111	0.729	0.562
6	SO_3	0.0276	0.0671	0.0274	0.624	0.0569	0.0799
7	K_2O	6.04	6.02	4.24	1.84	0.0812	11.8
8	CaO	0.993	1.84	2.32	46.8	6.98	3.15
9	TiO_2	0.122	0.189	0.654	0.320	6.13	0.366
10	MnO	0.0363	0.0440	0.0752	0.0296	0.982	0.0654
11	Fe_2O_3	1.19	1.42	3.91	0.106	0.215	3.28
12	ZnO	0.0062	0.0055	0.0083	3.03	7.34	0.0124
13	Ga_2O_3	0.0254	0.0253	0.0145	0.0160	0.0159	0.0175
14	Rb_2O	0.0116	0.0121	0.0404	0.0090	0.0095	0.0702
15	SrO	0.0069	0.0007	0.0001	0.0525	0.0348	0.0649
16	Y_2O_3	0.0031	0.0095	-	0.0012	0.0823	0.0000
17	ZrO_2	0.0123	0.0027	-	0.0142	0.0013	0.0223
18	BaO	-	-	-	-	0.0804	-
19	PbO	-	-	-	-	-	-

The results obtained from the powder X-ray phase analysis identify the phase composition of the samples of rock materials and building materials. The phase composition is also confirmed by the data from WD-XRF in Tables 1 and 2. In the rock samples the content of the main rock-forming chemical elements and their oxides is in the largest amount, for example SiO_2 - 68-76%. In the case of construction materials, the amount of variations is in a wider range - 38-71%. When using these materials for construction purposes, it is logical that the SiO_2 content is also high. Like SiO_2, Al_2O_3 varies in the narow range of 13-16%. The differences in the content of CaO are larger, as its quantity depends on the type of raw materials used for construction. In samples in which limestone or lime is added, the CaO content reaches up to 48%, and in rock and construction materials with high SiO_2 content, resp. in them the amount of CaO - decreases. It is important to note that the content of heavy metals in the presented samples is low and it does not pose a risk to human health.

The main risk to human health remains the content of radioactive elements that may be present in building materials from raw materials for their production. Therefore, gamma spectroscopy experiments were performed.

3.3. Gamma spectroscopy and the specific activities of ^{40}K, ^{226}Ra u ^{232}Th

To assess the potential of building materials from the studied areas, gamma-spectroscopic analyzes of samples of different materials were performed. The object of research are both natural rocks and interior materials such as old faience tiles and mortar, lime plaster and meta granite flooring, and basic materials for rough construction - concrete, concrete granite for cladding and flooring, gneiss, breccia conglomerates and granites. It is already known that due to the shorter half-life, the determination of the radiation intensity with respect to ^{40}K, ^{232}Th и ^{226}Ra is essential for the formation of radon immission levels indoors.

Based on the results of measuring the specific activities of ^{40}K, ^{226}Ra and ^{232}Th, an index of specific activity of building materials is calculated:

$$I = \frac{C_K}{3000\,\text{Bq/kg}} + \frac{C_{Ra}}{300\,\text{Bq/kg}} + \frac{C_{Th}}{200\,\text{Bq/kg}} \qquad (1)$$

where C_K, C_{Ra} and C_{Th} are the correspondingly measured specific activities of ^{40}K, ^{226}Ra и ^{232}Th. For assessment of conformity the maximum admissible values of the index of specific activity of construction materials, established by art. 5 of Directive №25 of 2005 on the requirements for protection of persons in case of chronic radiation as a result of production, trade and use of raw materials, products and goods with increased content of radionuclides.

Gamma-spectroscopic analyzes of samples of different materials were performed to assess the potential of the construction materials from the studied regions of the south-eastern and south-western directions in Bulgaria. The object of research are materials from the interior such as old faience tiles and mortar, lime plaster and meta granite flooring, and basic materials for rough construction - concrete, meta granite for cladding and flooring, gneiss, breccia conglomerates and granites.

The data obtained from the gamma-spectroscopic measurements of specific activities of ^{40}K, ^{226}Ra and ^{232}Th from scales and samples are presented in Table 3 and Table 4.

Table 3. Results of gamma spectroscopy of rock materials used in construction

№	**Objects**	**Material**	**^{40}K (Bq kg^{-1})**	**^{226}Ra (Bq kg^{-1})**	**^{232}Th (Bq kg^{-1})**
1.	Kardzhali city - PGCHE "Hristo Botev"	Gneiss	**1186±31**	**90.7±2.5**	**22.3±0.7**
2.	Zlatograd city - hotel	Deformed granitoid	**1259±33**	**94.0±2.3**	**27.8±0.9**
3.	Zlatograd city - hospital	Leucocratic metagranite flooring	**1212±32**	**58±8**	**49.0±2.1**
4.	Mineral baths villige, Haskovo - hotel	Metagranite wall cladding	**1176±30**	**77±6**	**31.1±1.2**
5.	Vinevo villige - house	Granitograis	**477±13**	**39.7±3.6**	**71.7±2.3**

All tested samples are below the maximum permissible concentrations of the requirements (Directive №25 of 2005) for the requirements for protection of persons in case of chronic radiation as a result of production, trade and use of raw materials, products and goods with increased content of radionuclides.

Table 4. Results of gamma spectroscopy of building materials

№	**Object**	**Material**	**^{40}K (Bq kg^{-1})**	**^{226}Ra (Bq kg^{-1})**	**^{232}Th (Bq kg^{-1})**
12	Kardzhali city - student dormitories	Old faience tiles	354±12	37.0±1.1	30.9±1.1
1	Haskovo - Mathematical High School "B. Petkanchin"	Old concrete	366±9	23.8±2.6	15.6±0.5
2	Haskovo - Sports Hall - chimney body	Old concrete	711±19	64±6	57.9±1.8
9	Haskovo city - museum	Old lime plaster	287±9	14.9±0.6	17.0±0.5
7	Haskovo - kindergarten "Bell"	Old mortar	630±16	34.9±3.3	28.1±1.1
6	Kardzhali - Pencho Slaveykov High School	Old mortar	740±20	23.9±0.8	35.5±1.4

8	Karamantsi village – "Hristo Botev" High School	Old mortar	757±21	36.7±1.4	57.2±2.1
13	Vinevo village - house	Sandstone	310±15	27.7±0.8	20.2±0.9

During the control of all construction materials from southeast direction in Bulgaria the following was established that index of specific activity of construction materials is in the range 0.13 ÷ 0.98 - below 1.00, the maximum allowable value of the index of specific activity of construction materials for the main construction of residential and public buildings.

4. Conclusions

The results of the analytical studies show that the presence of oxides of heavy metals, radionuclides in rock materials used for construction purposes and samples of different types of construction materials are below the maximum allowable concentrations, according to Bulgarian legislation for the maximum allowable index of specific activity for residential and public buildings. This may be due to the origin of the rock materials, and on the other hand to the fact that the surveyed buildings are more than 20 years old. Despite the results obtained, the risk in the construction of new buildings for the use of rock materials that generate radionuclides remains. It is necessary to develop a national policy for control of construction materials in order to prevent public health and reduce the risk to the public from exposure to radionuclides in indoor air.

5. References

[1] Mustonen R, Pennanen M, Annanmäki M and Oksanen E 1999 Enhanced Radioactivity of Building Materials. *Final report of the contract No 96-ET-003 for the European Commission. Radiation and Nuclear Safety Authority – STUK, Finland,* **96**

[2] Council Directive 96/29/EURATOM of 13 May 1996 laying down the basic safety standards of the health of workers and the general public against the danger arising from ionising radiation.

[3] Council Directive of 21 December 1988 on the approximation of laws, regulations and administrative provisions of the Member States related to construction products.

[4] *Powder Diffraction File* (PDF), copyright by International Centre for Diffraction Data (ICDD) Newtown Square; 12 Campus Blvd., Newtown Square, PA 19073-3273

[5] Duy Michael Le, Hanne R Sørensen, Anne S Meyer, 2017 *Biomass and Bioenergy* **97** 70e76, http://dx.doi.org/10.1016/j.biombioe.2016.12.018

[6] Meenakshi Singh, Singh M R , Sharma Anjali, Dighe Bhushan 2018 *Nuclear Inst. and Methods in Physics Research B* **436** 163–172, https://doi.org/10.1016/j.nimb.2018.09.022

[7] Erdtmann G, Soyka W 1979. The gamma of the radionuclide: tables for applied gamma ray spectrometry, *New Yourk:Verlag Chemic*

[8] IEC 61452:1995-09 *Nuclear instrumentation – Measurement of gamma-ray emission rates of radionuclides – Calibration and use of germanium spectrometers. International Electrotechnical Commission,* Geneva

Acknowledgements. This work was supported by the Operational Program "Science and Education for Intelligent Growth", co-financed by the European Union through the European Structural and Investment Funds under grant BG05M2OP001-1.001-0008 of National Centre for Mechatronics and Clean Technology (V.P.) and Bilateral project between Tallinn University of Technology of Estonian Academy Sciences and "I. Kostov" Institute of Mineralogy and Crystallography of Bulgarian Academy of Sciences (2022-2023) (V.P., E. S., T.K.).

Methodology for assessing of the restoration processes in reclaimed areas disturbed by polymetallic ore mining

V Stefanova[1] and P Petrov[1]

[1] Faculty of ecology and landscape architecture, Univertity of Forestry, Kliment Ohridski 10 Blv., 1756 Sofia, Bulgaria

E-mail: venistefanova3@gmail.com

Abstract. The present study develops a methodology for assessing the restoration processes in recultivated lands, disturbed by polymetallic ore mining. The methodology classifies the recultivated lands into four classes based on the recovery coefficient and the threat coefficient introduced through it. The coefficients are calculated for three groups of indicators - soils (formation of soil profile, mechanical composition, soil fertility), microflora (qualitative and quantitative composition), lack of threats (total pollution with heavy metals and metalloids, salinization, and acidification) relative to the studied site and a pre-selected reference site with similar characteristics.
Keywords: restoration, reclaimed, mining

1. Introduction

Technological progress is the reason for the continuous increase in demand for various metals - copper, lead, zinc, etc., and increasing intensive extraction and processing of polymetallic ores [1]. Mining activities are related to disturbance and pollution of the environment. In many cases, despite the post-liquidation activities on the area of the mines, which have exhausted their reserves, the impact on the components of the environment remains significant and is characterized by a large territorial scope and duration. The restoration of the areas disturbed by the extraction aims at restoring the ecosystem in the mining areas so that it is as close as possible to the natural ecosystem before the extraction in the area begins. This is essential both for the mining industry and for the protection of the environment and human health [2- 14].

Although environmental restoration is an extremely important process and there are several manuals, tools, etc. related to it, none of them describes the criteria for mine closure and assessment of restoration processes in ecosystems in these areas [15, 19]. The earliest assessment of the degree of restoration of disturbed areas and ecosystems are related to the assessment of plant species in reclaimed lands. Over time, the need to introduce "performance indicators" and the realization that they will be different for each area depending on the specific conditions increases [15]].

At present, general guidelines for assessing the restoration of damaged ecosystems, regardless of the reason for their degradation, are being introduced by the Society for Ecological Restoration through International Principles & Standards for the Practice of Ecological Restoration. Although the Standards can be applied to mining areas, they are not specifically targeted at them and do not provide information on the stages of recovery processes in this type of area [20]

All this necessitates the development of a methodology for assessment of the restoration processes, which will help to determine indicators for the restoration of these areas even before the reclamation.

It is also necessary to provide a quick and easy way to assess the stage of development of reclaimed land, as well as planning the necessary maintenance and corrective activities of reclaimed land.

The aim of the present study is to develop a methodology for assessing the ongoing recovery processes based on selected indicators.

2. Reasons for developing the methodology. Purpose of the methodology

2.1. Reasons for developing the methodology

The main motives for the development of the Methodology for assessment of the restoration processes in reclaimed areas are related to the need to select specific indicators that are important for the restoration of soils and the ecosystem. It is also necessary:

- quantitative measurement of the restoration in the disturbed areas;
- quick and easy way to assess the stage of development of reclaimed area;
- planning the necessary maintenance and corrective activities in this type of area;
- planning the necessary additional reclamation;

2.2. Purpose of the methodology

The development of the methodology is in line with four main purposes:

- oopportunity for comparative analysis of ongoing recovery processes.
- quantification of the progress regarding the restoration of the reclaimed area.
- providing an opportunity for fast and reliable assessment.
- development of appropriate scales to provide reliable information on the degree of recovery.

3. Main definition

For the purposes of the Methodology, the following basic concepts have been introduced:

- *Reference site (reference ecosystem)* - The SER standard recommends the use of a reference site and comparison of the compliance of the status of pre-defined groups of indicators in it and the ecosystem subject to study.
  In this case, the reference site is an ecosystem located near the studied site, which has similar features and is exposed to similar environmental factors [19] and impact factors relevant to soil formation, for example: climate, relief (exposure, slope), soil-forming rock, biotic soil-forming factors - vegetation, animal organisms and microorganisms, etc. The purpose of the use of a reference site is to clearly define the purpose of restoration and to quantify the ongoing processes in the reclaimed land, compared to the reference. In this regard, the choice of reference object is extremely important to obtain reliable results [19]
- *Object of research* - Reclaimed land disturbed by the extraction of polymetallic ores, subject of research and evaluation.
- *Recovery indicator* - An indicator that is relevant to the restoration of areas and serves to quantify the degree of restoration of disturbed area [19]. In the present case, for example, the formation of soil profile, mechanical composition, organic matter content, etc.
- *Groups of indicators* - For the purposes of this methodology, the grouping of indicators is done in three groups Soils, Microflora, No threats. Each group of indicators includes complementary indicators and provides a reliable assessment for the respective group.
  - Indicator group "Soils" - includes basic indicators of soil properties.
  - Indicator group "Microflora" - includes basic indicators of soil microflora.
  - Indicator group "Lack of threats" - includes major pollutants relevant to the specific area. The indicators for this group of indicators are determined depending on the origin of the recovery area and for each specific site.

The individual indicators for each group, as well as the reasons for their inclusion in the respective group are presented in Table 1.

Table 1. Indicators for assessment of restoration processes in reclaimed areas

Indicator groups	Indicator	Motive for inclusion
Soil	Formation of soil profile (profile depth)	The result of soil formation is the development of the soil profile. The formation of the soil profile provides information about the degree of soil formation process [20].
	Layer formation of Leaf litter (layer thickness)	Initiates the stage of organic matter accumulation and the initial stages of humus formation [21].
	Content of clay fraction	It is associated with the process of clay formation, which is part of the elementary processes of soil formation [22].
	Contents of org. C / Humus content	The formation of humus is a sign of the last stage of soil formation [21].
	Content of total N	Nitrogen is a major nutritional and limiting factor in plant development [21].
	P_2O_5 content	Phosphorus is a major nutritional and limiting factor in plant development [23].
	K_2O content	Potassium is a major nutrient and limiting factor in plant development [21].
Microflora	General microflora	It is related to the processes of decomposition of organic matter and humus formation [21, 22].
	Bacilli	A sign of the occurrence of soil self-cleaning processes [24, 25].
	Non-spore-forming bacteria	Participate in the processes of decomposition of organic matter in precipitation [26]
	Mold fungi	They transform organic matter into forms that other organisms can absorb [27, 28].
	Actinomycetes	Participate in nitrogen fixation and degradation of high molecular weight compounds such as hydrocarbons, improve nutrient availability, participate in the formation of stable humus, and combine with other soil microorganisms in the degradation of solid plant residues such as cellulose and animal residues, participate in the cycle of substances [29].
	Bacteria assimilating mineral nitrogen	Participate in the processes of transformation of nitrogenous substances in soils and nitrogen nutrition [24].
	Mineralization coefficient	A sign of the activity of microorganisms and an assessment of the degree of mineralization processes [24, 25].
Lack of treats	Contamination with heavy metals and metalloids	The presence of contamination with heavy metals and metalloids leads to a delay in the recovery process - inhibition of vegetation development and can affect the activity of microorganisms. Expected contamination with heavy metals and metalloids due to increased lithogeochemical background [13, 21]
	Salinization	Reduces soil fertility and inhibits vegetation development. Leads to physiological drought [13, 21]
	Acidification	Inhibits the development of vegetation and leads to the mobilization of heavy metals and metalloids [13, 21, 22]

- *Test site* - Part of the object of study, which is characterized by certain conditions.
- *Test point* - A point on the reference site where sampling is performed, or data are available on the condition of the soils therein.
- *Reference value* - Mean value with standard deviation obtained for an indicator in the reference site or the studied site after statistical processing of the data on its condition.

4. Basic steps for assessing the restoration of disturbed area

4.1. Selection of a reference site

To obtain a correct assessment when applying the methodology, it is essential to correctly determine the reference site. The reference site must correspond to the type of ecosystem to be restored and subject to the same influencing factors.

It is necessary to clearly identify the type of ecosystem that has been disturbed, damaged, or destroyed, such as forest ecosystem, grassland ecosystem, agroecosystem, etc. Example: if disturbed area is restored as a forest area, it is necessary for the reference site to be a forest area, as close as possible to the species composition to the species used for reclamation.

The choice of the reference site must be motivated by all the indicators listed below. The information received about its state must be reliable and give a clear idea of the state of the ecosystem in it. The selection of a reference site should be made after an in-depth environmental assessment of the area, considering all key components and environmental factors.

The purpose of the selected reference site is to provide information on the expected result of the restoration of the reclaimed area and the stage at which the reclaimed land is located, in terms of restoration. Therefore, the closer the values of the assessment indicators between the reference site and those of the reclaimed area, the more advanced the restoration of the area.

For the assessment of the ongoing restoration processes in the reclaimed area to be reliable, the selected reference site should correspond to the recultivated one, at least according to the following indications:

- Location and relief (altitude, exposure, slope) [30];
- Environmental factors (climate, hydrological regime, pollution) [30];
- Soil-forming rock [13];
- Type of ecosystem [19].

These indicators are directly related to and influence the formation of the soil layer and the course of the soil formation process.

It is recommended to select at least 3 test points, data for which to be used in the application of the Methodology, to ensure the reliability of the information obtained on the condition of the reference site.

4.2. Data collection and statistical processing

Data collection can be done in two ways - by own sampling according to the requirements of BDS EN ISO 22475-1: 2022 or by databases - monitoring of the Companies, large-scale monitoring, or other databases with guaranteed reliability.

The information collected must be processed to obtain a reference value for the reference site - a mean value with standard deviation. The stages of collection and processing are presented in Figure 1.

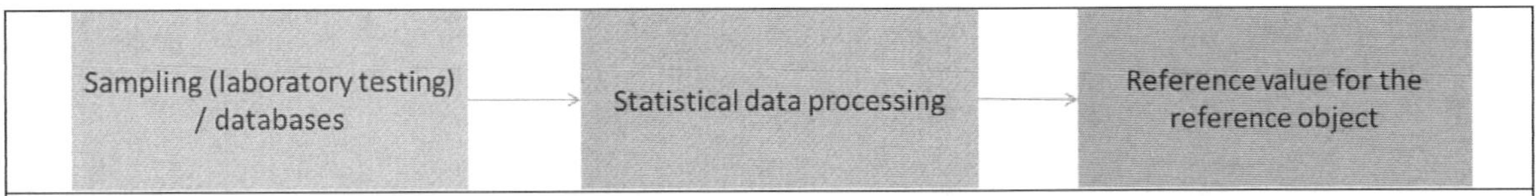

Figure 1. Stages of collection and processing of data

Statistical data processing includes the elimination of gross and system errors, calculation of the average values of the data set and standard deviation.

4.3. Calculation of recovery and threat ratios

For the purposes of the Methodology, two coefficients have been developed - recovery coefficient and threat coefficient.

Assessment is performed for each of the indicators in the three groups separately for both layers - surface and subsurface. To assess the recovery, use the reference value obtained for the reference site for each indicator and calculate the recovery factor. The recovery coefficient reflects the ratio between the value of soil indicator / microflora indicator of the reclaimed land compared to the reference value of the same parameter in the reference sites.

$$\text{recovery coefficient} = \frac{\text{value of the indicator in the reclaimed area}}{\text{reference value for the reference site}}$$

The maximum result that can be obtained by applying the formula is 1. If any of the indicators has a higher (more favourable) value in the samples of reclaimed land compared to the reference site, the value obtained is equal to 1. The minimum the result is - 0.

The formula is applicable to the parameters included in group 1: Soil and group 2: Microflora.

The recovery rate is not applicable for group 3: Lack of threats. It is possible due to the natural conditions in the area (geophone), in the reference site to be observed deterioration of the relevant parameters - heavy metal pollution, salinization and acidification. For this purpose, a threat coefficient is introduced. For its calculation, values for the reference site are not used, but values according to normatively set MPCs or reference scales, as follows:

- for the indicator "content of heavy metals and metalloids" the maximum admissible value according to Ordinance 3/2008 or data for the background content of the respective element is used.
- the value 4 mS.cm-1 is used for the "salinity" indicator, as below this value the studied soils are considered as unsalted.
- the indicator "acidification" for the target (optimal) value is 6.0, because below this value most heavy metals and metalloids increase their mobility and increase the risk to the environment [23].

The calculations for the parameters in the third group are performed as follows:

$$\text{threat coefficient} = 1 - \frac{\text{value of the indicator in the reclaimed area}}{\text{value according to normative document / scale}}$$

The maximum result that can be obtained by applying the formula is 1, and the minimum - 0. In case the value of the coefficient exceeds 1, it is equated to one, and lower than 0 values are equated to zero. The closer the value is to 0, the bigger it is the threat on the respective indicator.

The elements included in the heavy metals and metalloids pollution indicator may differ depending on the expected pollution. Since the tested elements are more than 1, the lowest obtained coefficient is taken, the one that considers the greatest threat of pollution.

4.4. Filling in the recovery card

The obtained results are filled in a card of the recovery, which gives a visual idea of the ongoing recovery process - Figure 2.

Rcovery card						
Site		....				
Indicator		Soil layer	C I (0.1)	C II (0.2-0.4)	C III (0.5-0.8)	C IV (0.9-1.0)
Soil	Soil profile					
	Leaf litter					
	Clay content					
	Humus content	surface				
		subsurface				
	Content of Total N	surface				
		subsurface				
	P_2O_5 content	surface				
		subsurface				
	K_2O content	surface				
		subsurface				
Microflora	General microflora	surface				
		subsurface				
	Bacilli	surface				
		subsurface				
	Non-spore-forming bacteria	surface				
		subsurface				
	Mold fungi	surface				
		subsurface				
	Actinomycetes	surface				
		subsurface				
	Bacteria assimilating mineral nitrogen	surface				
		subsurface				
	Mineralization coefficient	surface				
		subsurface				
lack of treats	Contamination with heavy metals and metalloids	surface				
		subsurface				
	Salinization	surface				
		subsurface				
	Acidification	surface				
		subsurface				
Total number of restored parameters						
Recovery category						

Figure 2. Recovery card

For each indicator, the value obtained for the respective coefficient - recovery and threat to the surface and subsurface soil layer - is entered in the card in four classes as follows:

- Class I (C I) - The coefficient value up to 0.10;
- Class II (C II) - The coefficient value 0.20 - 0.40;
- Class III (C III) - The coefficient value 0.50 - 0.80;
- Class IV (C IV) - The coefficient with value 0.90 - 1.00.

When the coefficient has a value in Class 1, its order is coloured only in Class 1. If the coefficient has a value in Class 2, then the order in Class 1 and Class 2 is coloured.

Filling in the card in this way gives the opportunity to make an easy visual assessment of the degree of recovery.

4.5. Scale for assessing the degree of recovery

Based on the number of matches identified through the recovery card, the disturbed sites are classified into four classes, describing the stage of each ecosystem in the overall recovery cycle (Table 2).

Table 2. Four-point scale for assessing the degree of restoration of reclaimed land using the restoration card

Class	Characteristics of the recovery
Class I - Disturbed area	The value of 70% and more of the coefficients for the indicators of man-made soils is below 0.10 in all groups. The area is considered to have been disturbed and reclamation activities need to be undertaken. In case the area has already been recultivated, it is considered that restoration processes have not taken place yet - it is necessary to repeat the assessment after 1 year and, if necessary, to take corrective actions.
Class II - Initial stage of restoration	The value of 70% and more of the coefficients for the indicators of man-made soils is in the range 0.20 - 0.40 in all groups.
Class III- Advanced stage of restoration	It is considered that the initial stages of recovery are beginning. Corrective action may be needed to speed up the recovery process.
Class IV - Restored area	The value of 70% and more of the coefficients for the indicators of man-made soils is in the range 0.50 - 0.80 in all groups.

5. Conclusion

The methodology is applicable only for reclaimed terrains disturbed by the extraction of polymetallic ores. For its application in the extraction of coal or non-metallic minerals, it is necessary to rework a part „Lack of treats".

6. References

[1] Miao, Z, R Marrs 2000. Ecological restoration and land reclamation in open-castmines in Shanxi Province, China, *Journal of. Environmental Management*. (59) 205-2015.

[2] Kostadinova, A, E Todorova, T Krumuv, Opportunities for stabilization of forest roads using waste from coal mining, *Physics and mathematics*, ISSN 2518-167X

[3] Todorova, E, A Kostadinova 2019. Risks of environmental pollution from mining waste from ore-containing copper processing, *Journal of environmental protection and ecology* 20(1):397-403

[4] Zukova-Aleksandrova D., S. Mitkov 2018. Geological, climate and anthropogenic prerequisite for proceeding of erosion processes on the territory of Madzharovo caldera, Ecology & Safety ISSN 1314-7234, Volume 12, 2018

[5] Zukova-Aleksandrova D, S Mitkov 2019. Artificial forest ecosystem on the coal mine dumps of Maritza Iztok mines in Bulgaria, *Journal of Balkan Ecology*, Volume 22, No 4

[6] Bancheva-Preslavska, H 2017a Framework for erosion control measures in protected areas, *Web of Scholar*, 9(18), Vol.1, ISSN 2518-167X

[7] Bancheva-Preslavska, H 2017b Forest amelioration for erosion control in nature park, Vrachansky balkan "through volunteer activities, *Наука – кн.* 1/2018, том XXVIII Издание на Съюза на учените в България

[8] Bezlova, D, M Doncheva, E Tsvetkova. K Vasilev 2016 Survey of the concentration of

heavy metals in beech leaves in the region of Central Balkan National Park, Bulgaria, Phytologia Balcanica 22(3):335-339

[9] Павлов, П 2018 Въздействие на минералносуровинния отрасъл върху земите и почвите. Последващо устойчиво управление. Минно дело 19-23 ISSN 0861-5713.

[10] Павлов, П 2016 Нарушения на почвената покривка от промишлени и минни дейности, и методи за възстановяване. София: *МГУ "Св. Иван Рилски"*.

[11] Павлов, П, Л Толев 2010 Воздействие добычи руды на баланс экосистем. Москва. *Научно-технический журнал* №2. 50-62.

[12] Желева - Богданова Е 2010. Рекултивация на нарушени терени, София: PSSE.

[13] Петрова, Р, Л Тотев, П Пвалов, С Анисимова 2017 Рекултивация на терени, нарушени от добива и преработката на полезни изкопаеми. София: *МГУ "Св. Иван Рилски"*.

[14] Blommerde, M, R Taplin, S Raval 2015 Assessment of Rehabilitation Completion Criteria for Mine Closure Evaluation, (Sustainable Development in the Minerals Industry).

[15] Buxton, A. 2002. Mining, Minerals and Sustainable Development (MMSD). IIED.

[16] Cairns, Jr. 2000. Setting ecological restoration goals for technical feasibility and scientificvalidity, *Ecology Engineering* 15 171–180.

[17] ICMM 2006 Guidance Paper Financial Assurance for Mine Closure and Reclamation, International Council on Mining Metals.

[18] ICMM 2011 Planning for Integrated Mine Closure: Toolkit, London: International Council on Mining Metals, London

[19] McDonald, S, S M Landhäusser, J Skousen, J Franklin, J Frouz, S Hall, D F Jacobs S Quideau 2015 Forest restoration following surface mining disturbance: challenges and solutions. *New Forests volume* 46 703–732.

[20] Ankush, J 2018 How Soils are Formed: Stages, Process and Factors | Soil Science, https://www.soilmanagementindia.com/soil-formation/how-soils-are-formed-stages-process-and-factors-soil-science/15711.

[21] Dinheiro, V., N. Nhamo, G. Kyalo. 2014. Exploring Options for Lowland Rice Intensification under Rain-fed and Irrigated Ecologies in East and Southern Africa: The Potential Application of Integrated Soil Fertility Management Principles - ScienceDirect (article.dyndns.info), *Advances in Agronomy* https://doi.org/10.1016/B978-0-12-802139-2.00005-6.

[22] Малинова, Л 2010 Почвознание и замърсяване на почвите, София: *Издателска къща ЛТУ*

[23] Daroub, S, A Gerakis, J T. Ritchie, D K Friesen, J Ryan 2003 Development of a soil-plant phosphorus simulation model for calcareous and weathered tropical soils, *Agricultural Systems*, 2003, vol. 76, issue 3, 1157-1181

[24] Малчева, Б 2020 Почвено - микробиологични показатели за установяване на статуса на антропогенни почви на територията на община София, Варна: *ИК Геа - Принт*.

[25] Наскова, П, М Консулова, Д Пламенов, Б Малчева. 2017 Математически модел за определяне на степента на влияние на различни физико-химични фактори върху числеността на общата микрофлора в антропогенни почви, *Ново знание*, 6-4.

[26] Григорова-Пешева Б 2019 Влияние на факторите на околната среда върху микробната ценоза при два типа горски почви *Почвознание агрохимия и екология*, 53, 34-44.

[27] Ingham, E 2017 The living soil: fungi, United States Department of Agriculture, https://www.nrcs.usda.gov/wps/portal/nrcs/detailfull/soils/health/biology/?cid=nrcs142p2_053864.

[28] Ingram, L, J Stahl , G E Schuman, J Buyer, G F Vance, G K Ganjegunte, J W Welker 2008 Grazing affects on carbon, nitrogen and microbial communities in a mixed grassland, *Soil Science Society of America Journal* 72: 939–948. DOI: 10.2136/sssaj2007.0038.

[29] Bhatti, A, S Haq, R Bhat 2017 Actinomycetes benefaction role in soil and plant health *Microbial Pathogenesis*, 111, doi:10.1016/j.micpath.2017.09.036 458–467. .

[30] Петров, П 2019. Подходи за рекултивация на Източно насипище, рудник "Елаците", София: *Авангард Прима*.

IOP Conf. Series: Materials Science and Engineering 1264 (2022) 012005 doi:10.1088/1757-899X/1264/1/012005

Potential of phytomining in Bulgaria

Iva Miteva[1], Petar Petrov[1], Veneta Stefanova[1]

[1]University of Forestry, Sofia, Bulgaria

E-mail: iva.i.miteva@gmail.bg, petargpetrov@abv.bg, venistefanova3@gmail.com

Abstract: Phytomining has been gaining more and more popularity in recent years. Scientific information is being accumulated and technologies are being developed that could make phytomining an applicable technology for the extraction of rare and noble metals.
In Bulgaria, the mining industry is well developed. Copper, lead, zinc, gold and polymetallic ores are mainly mined. The waste generated during the mining process of these ores has a great potential for phytomining. There are many existing, abandoned and closed dumps, tailings ponds and landfills that have the potential to be used for phytomining.
This paper analyzes the possibility of using phytomining technology for the extraction of noble and rare metals in Bulgaria. An overview of the level of development of phytomining technologies and the available resources in Bulgaria for which it is applicable has been made.

1. Introduction

The development of innovative approaches for obtaining prices and rare metals from waste is a topical and significant issue in the scientific community and business. The importance of development is determined by the high value of these metals as a strategic raw material and as a material with extremely important application in the development of new technologies. The implementation of the actions and projects is of great importance to the circular economy and will contribute to the extraction of previously inaccessible EU resources, including in Bulgaria.

Hundreds of millions of tonnes mining waste from active or abandoned mines have been disposed close to the local population without treatment, because of extraction and processing of minerals. During the exploration of mines, the available technology at the time used to concentrate target minerals often is not as good as those currently available. Therefore, the waste materials from metal mining often contain significant amounts of potentially valuable metals. The amount of waste produced per unit of ore will increase in the coming years as a result of exploitation of increasingly impoverish deposits [1-3, 25].

Phytomining is an innovative way to recover metals from soils, and it has attracted more and more attention in recent decades [4]. A scientometric review of the published literature on the issue between 1972 and 2022 shows that the Web of Science published more than 232 publications on phytoproduction, and the number of publications maintains a drastic increase each year, reflecting global interest in this topic. There are still a number of uncertainties related to the mechanisms of accumulation of noble and trace metals and determining the potential for their phytoproduction [5].

Bulgaria falls under the definition of a "mining country" with metal extraction of 11 tons per person. Among the most mined metals are copper, lead, zinc, and gold. In relation to the mining operations within the country, according to the data of the Ministry of Environment and Waters, by

IOP Conf. Series: Materials Science and Engineering 1264 (2022) 012005 doi:10.1088/1757-899X/1264/1/012005

2022, the total number of tailings repositories in Bulgaria is 33. 21 have been decommissioned and recultivated and 12 are in operation [6, 7].

The aim of the present study is to examine the potential of phytomining of noble and rare metals from landfilled waste from the mining industry in Bulgaria.

2. Methods of work

To achieve the objectives of this paper, the method of cross-sectional research is applied. It includes a combination of methods that allow data analysed by cross-checking from a different number of sources. These include: regulatory documents; scientific articles; monographs; expert reports, statements, opinions, etc. by national and international experience. As a result, more than 305 Mt of the amount of mining waste from closed tailings and over 680 Mt from existing tailings were studied. The appropriate vegetations and the applicable ameliorants were determined based on the observed literature for the tailings chemical composition.

3. Results

3.1 Deposits, landfills, tailings, and embankments in Bulgaria

On the territory of the country there are nine defined ore regions - 3 for copper and 5 for zinc-lead polymer ores. Copper ores are of great importance for the development of Bulgarian non-ferrous metallurgy. They have a low metal content (about 0.5% to 2%). Their main deposits are in the Balkan Mountains and small quantities of copper ore are mined in the Burgas region. There are about 46 polymetallic deposits in the Rhodope ore zone and their origin is related to volcanic activity during the Oligocene [9].

Figure 1. Map of Bulgaria with tailing pounds

Panagyurishte Ore Region

The Panagyurishte Ore Region is an element of the Apuseni-Banat-Timok-Srednogorie Magmatic and Metallogenic Belt and is the most important source of copper and gold in Bulgaria [8]. The mining activity in this region includes Asarel Medet, Elatsite med, Elshitsa, Medet, Vlaykov Vruh, Tsar Asen, Chelopech, Radka, Krasen-Petelovo. Krasen-Petelovo is a non-industrial gold deposit [11]. Gold and precious metals in the copper deposits of the region are deposited during the main stages of hydrothermal mineral formation [12]. In Elshishko - Radinsko ore field there are three types of ore

IOP Conf. Series: Materials Science and Engineering 1264 (2022) 012005 doi:10.1088/1757-899X/1264/1/012005

bodies: pyrite, copper-pyrite and copper-pyrite-polymetallic. The most common minerals are quartz, pyrite, chalcopyrite, cubanite and others. [11]. According to Popov et. all districts have been exploited mostly for Cu-Au, S [12]. The gold content in pyrite and chalcopyrite for deposits Elshitsa, Radka, Krasen is from 0.4 to 3 g/t; for Chelopech deposit from 1 to 5 g/t; for Tsar Asen, Assarel, Medet deposits is from trace to 1 g/t and from trace up to 9 g/t for Ellatsite deposit. The gold contents in bornite vary from 1 to 50 g/t for the copper-pyrite type of deposits and from traces up to 20 g/t for the Ellatsite deposit. Chelopech mining complex includes extraction of copper-gold-pyrite ores from the Chelopech deposit [13] with gold content is from 5 to 20 g / t [14] in tenanite and enargite. In Elacite Med the copper, cadmium and lead have the highest concentration above the maximum allowable concentrations. The amounts of zinc, cobalt and chromium are smaller, and all samples contain arsenic [15].

Vratsa ore region

There are four processing plants with tailing in Vratsa ore region- Martinovo, Chiprovtsi and Eliseina and Mir. In the Martinovo deposit iron is the main metal therefore copper, lead, zinc, and gold can be mentioned as secondary [12].

In the Chiprovtsi deposit more than 40 minerals have been described, excluding the supergenic ones, which form several mineral associations - iron carbonate (siderite), iron sulphide, silver - lead (polymetallic), quartz-calcite-barite-fluorite and pyrite-cinnabarite. Native Au, Ag and Sb have been found in the deposit. Of the greatest industrial importance are galena, chalcopyrite, arsenopyrite, quartz, calcite, fluorite, anchorite, ferrodolomite. Many supergenic minerals have been identified, such as malachite, azurite, cerussite, manganese and iron hydroxides, covelline, etc., which have a limited distribution [16].

The ore in the tailing pound Mir has two types of sulphide mineralizations: lead-zinc and lead-copper-silver with bismuth [17]. On May 1, 1966, the tailings pond of the zinc and lead collapsed leading to huge devastation [18]. Due to the accident the contamination is spread on a long distance and would be hard for localization.

To sum up the plant hyperaccumulators for Panagurishte and Vratsa ore regions have to be resistant to low pH and to accumulate Au, Co, Cu. There is a short data for available noble metals as Ge, Ni, Zn, Pt, Re but further investigations should be performed so to be estimated the quantity.

Burgas ore region

The second largest copper region in the country is the Burgas ore region with 4 main ore fields (from east to west - Rosen, Vurli breg, Zidarovo and Bakadzhik) in the same direction increases the relative share of polymetallic mineralization at the expense of copper and gold concentration, and decreases of molybdenite and iron oxides (mainly magnetite). Rosen, Vurli Bryag, Zidarovo and Bakandzhiyskoto ore fields belong to this region [12]. The most common minerals are quartz, magnetite, pyrite, chalcopyrite, sphalerite, galena, malachite, azurite, limonite, chalcocite, cuprite, covelin and others. [19]. Despite the similarities, the deposits from the different ore fields have some differences:

- Rosen, chalcopyrite-molybdenite (cobalt-bearing) type - with the highest concentration of Cu, Co, Mo, V, U
- Vurli Bryag, chalcopyrite-speculator type - with the highest concentration of Cu, Bi, Pb, Zn
- Zidarovo and Balkandzhiysko - chalcopyrite-gold-polymetallic (bismuth-bearing) type - with the highest concentration of Cu, Bi, Au, Ag, Pb, Zn [11]

Malko Tarnovo ore region

In the area around Malko Tarnovo ore region can be found skarn copper-molybdenum-iron ore deposits - Gradishte, Propada, Mladenovo and Bardtseto. The ore for the whole region is processed in the Malko Tarnovo Processing Complex and the tailings are deposited in the tailings pond, currently remediated. The following mineral types of ores are distinguished: magnetite, magnetite-chalcopyrite,

chalcopyrite-specularite, chalcopyrite, bornite, chalcopyrite-bornite-scheelite and transition ores between them. The main sulphide minerals are chalcopyrite and pyrite, bornite, bismuthite, molybdenite, pyrrhotite, marcasite, tenanite, chalcocyte, and for the Bobertseto and Propada deposits both sphalerite and galena. Some gold-silver and nickel-cobalt minerals and sulfosols are rare, such as aikinite, dognachkaite, cuprobismutinite, galenobismutinite, vitihenite, gunahuatite, minegenite, tetradimite and others [11]. Malko Tarnovo Tailings Reservoir has been reclaimed.

Pursuant to Decree 140/1992 on the restructuring of ore mining and the phasing out of inefficient production facilities, all four processing plants from Burgas and Malko Tarnovo re fields were liquidated and remediated, together with the tailings ponds attached to them, which could be an obstacle for phytomining. Considering the location of the tailing the estimated species must be thermophilic and acidophilic and at the same time to have a high uptake of Cu, Mo, Bi, Pb and Zn. In Zidarovo, Balkandzhiysko and Malko Tarnovo could be appropriate to have Au hyperaccumulator plant species.

Eastern Rhodope ore region

On the territory of the eastern Rhodope there are there Processing Complex - Gorubso Kardzhali, Madjarovo and Ustrem. Kardzhali is dominated by Au-Pb-Zn and Au-hydrothermal degenerations. Anomalies of Au, Ag, Cu, Pb, Zn, etc. are also found in this area. It should be noted that most gold and gold-polymetallic deposits (Chala, Madzharovo, etc.) are processed by OF "Gorubso Kardzhali" AD and deposited at the tailings Kardzhali 2 [20].

In Madjarovo ore field have been found more than 170 minerals and mineral varieties in the in the host rocks. The minerals galena, sphalerite, chalcopyrite and gold are of the greatest importance. Most of the gold is in its unrefined form and only a small part is like aurostibit. Secondary, but also in large quantities are the minerals chalcopyrite, pyrite, tetrahedrite, hematite, pyrolusite, goethite, vavelite, pyromorphite, lead-antimony and silver-antimony sulfosalts. The presence of Ag, Sb, Bi, Te, Sb, Ga, Ge has been found in most of the minerals as impurities. [17]

The main ore minerals in the Ustrem deposit are galena, with a high content of Ag, and sphalerite, with a low content of Fe. All other minerals are secondary, with the largest amounts being pyrite, chalcopyrite, pyrrhotite, tennantite arsenopyrite, marcasite cerussite, anglesite pyromorphite barite, sericite, and others. [17]. The tailings pond has been reclaimed.

Western Rhodope ore region

The lead-zinc core and metasomatic type of the Madan ore region are one of the largest ore concentrations of these metals in Europe. [21]. The ore from Madan ore field is processed in two processing complex - Madan AD which has two tailings Rudozem and Rudozem-2, and Erma Reka AD, with its tailings. There have been 153 minerals and mineral varieties identified in the lead-zinc deposits. in the Madan ore field. They are being processed by Gorubso-Rudozem. The main minerals in the Madan ore field are galena, sphalerite, hydrogyote, goethite, quartz, johansenite. Secondary minerals are pyrite, chalcopyrite, arsenopyrite, manganocalcite, rhodochrosite, calcium ferromangandolomite, diopside, prolusite, cerussite, anglesite, pyromorphite and others. Although not in large quantities, there are also minerals from the tetrahedrite-tennantite series. The main impurity elements are Ag, Sb, Bi, Te, In, Co, Ga, Se, Ni [22].

Central Rhodope ore region

The Laky is Pb-Zn deposit with a lot of inclusions like pyrite and chalcopyrite and heavy metals [23]. The selected species for all Rhodope ore region must be resistant to moist and temperate climate, low pH and to be able to accumulate high concentrations of Au, Pb, Zn and Cu. Because the deposits are polymetallic it is worthy to be analyzed for noble metals as Ge, Pt, Ni, Co, and Re.

Phosphogypsum deposit near Dimitrovgrad
Phosphogypsum is the main by-product of the production of phosphoric acid and it is formed by acidic digestion of phosphate rock. The reserves of the metal-bearing potential of the phosphogypsum deposit amount to 8Mt, with a content of rare and scattered elements 40 Kt, strontium 140 Kt and titanium 24.6 Kt [24].

3.2 Noble metals available in Bulgaria and hyperaccumulator plants
Based on the conducted research, gold has the greatest potential for phytomining in Bulgaria. If a plant accumulates more than 1 mg kg^{-1} of gold, then it is considered as hyperaccumulator. This limit is based on the normal gold concentration in plants of 0,01 mg kg^{-1} [28]. There are scientific papers for more than fifteen hyperaccumulator species - *Brassica juncea* (Indian mustard), carrot, red beet, onion, radish[29], *Berkheya coddii* (Asteraceae), *Cichorium intybus* (chicory)[30], *Lupinus* sp. (blue lupin)[31], *Zea mays*(corn)[32], *Helianthus annuus L.* (sunflower), *Kalanchoe serrata L.* (magic tower)[27], *Lindernia crustacea* (Scrophulariaceae), *Paspalum conjugatum* (carabao grass), *Cyperus kyllingia* (nut grass)[33] Tobacco[34], *Brassica napus*[35]. From them *Brassica juncea*, *Berkheya coddii, Brassica napus*, *Helianthus annuus L. and Kalanchoe serrata* have the highest Au concentration and are suitable for planting in Bulgaria. *Brassica napus, Helianthus annuus L. and Kalanchoe serrata L.* shows also a good accumulation of Cu together with Au and are suitable for planting in tailing pounds in Panagurishte ore region, Vratsa ore region, Zidarovo, Balkandzhiysko and Malko Tarnovo, with the flowering temperature between 25.4°C - 3.9°C [36]. Tobacco also shows good uptake and is widespread in southern Bulgaria as agriculture therefore in Rhodopes. It is advantage that the local population have experience for growing it.

Rhenium is one of the rarest (7 x 10^{-8}%) and most broadly scattered elements on Earth's crust [37]. Due to its paucity (approximately five times scarcer than gold) [38], Re is one of the most valuable metals in the international market. Re is a by-product by molybdenum extraction which can be a by-product by the copper extraction [39]. By the vicinity of the Assarel-Medet copper mine (Panagurishte ore region). When research was carried out on alfalfa, it was found that the content of Re in plant leaves can reach up to 3985 mg kg-1, and in white clover up to 13 mg kg-1 [40]. Also, her discover is that Re is water soluble and could be leachate more than 99% by the vegetation. For Alfalfa that could have a double benefit– it would recover noble metal and could be used as agriculture.

More than 520 Ni hyperaccumulator species are currently known and some of these grow quickly providing a high-farming yield [41]. By the performed investigations only *Alyssum murale, Alyssum corsicum* [42] and *Berkheya coddii*[43] are suitable for growing in Bulgaria. *Alyssum murale, Alyssum corsicum* are widespread in whole Bulgaria and shows good uptake of Co together with Ni (*A. murale*: Co 52.4 mg kg^{-1}, Ni 6 080 mg kg^{-1}, *A. corsicum*: Co 48.7 mg kg^{-1}, Ni 4 940 mg kg^{-1}). These plants can be naturally found in sandy and stony meadows or in the foothills and mountainsand prefer neutral to high pH. Almost all of the pointed tailings in Bulgaria have low pH due to pyrite and other oxide minerals. Therefore, the Ph should be corrected whereas in the tailings is planted. In this line of thinking Phosphogypsum deposit near Dimitrovgrad would be appropriate for substrate for planting a. *murale and A.corsicum.*

Already known for gold accumulation *Berkheya coddii* is suggested as a Ni and Pt hyperaccumulator plant with rapid biomass production that can accumulate more than 14 000 mg kg^{-1} Ni and 0,5 mg kg^{-1}Pt [44]. The second investigation for Pt hyperaccumulator is made with *Sinapis alba* which is widely spread in whole Bulgaria. This plant can uptake more than 760 mg kg^{-1} Pt [45].

Cobalt and Germanium are critical raw metals in EU's list of Critical Raw Materials (2020). The analyzed data show that they are available in Eastern Rhodopes, Central Rhodopes, Vratsa and Panagurishte ore region. Germanium is mostly recovered as a by-product from zinc mines. [46] More than 40 plant species have been observed in scientific papers but only *Oryza sativa* (rise), *Triticum aestivum* (common wheat) [47], *Allium sativum* (garlic), *Solanum tuberosum* (potato)[48] and *Agrostis capillaris* [49] meet the conditions of Bulgaria and can accumulate huge amount of Ge. The highest concentration of Ge is found in Oryza sativa. *Agrostis capillaris* and *Oryza sativa* are native to Eurasia and prefer moist grasslands and open meadows. It is important to note that *A. capillaris* accumulate Ge in roots almost twice as in shoots. Because *Oryza sativa*, *Triticum aestivum*, *Allium sativum*, Solanum tuberosum are one of the most common agricultures in Bulgaria it would be hard to save the harvest by the local people, especially in non-guarded tailings.

Usually, Zn mines are differed also with high concentration of Pb. According to the global database for plants that hyperaccumulate metal and metalloid trace elements [50] for Europe and China there are *Arabidopsis halleri*, *Noccaea* and *Sedum plants* which can accumulate more than > 3000 mg kg^{-1}Zn and *Noccaea plant* which can accumulate more than > 1000 mg kg^{-1} Pb. *N. caerulescens* prefer high humidity and temperatures around 20°C so that proper conditions for planting it have in Rhodope ore regions. Sedum are succulent plants which means that they can survive in hot temperature and low content of water. Therefore, will be suitable for planting it in Vratsa, Panagurishte and Burgas ore regions.

4. Conclusion

With the evolution of technology, the humanity will need more and more resources. That will force the scientist to search for alternative variants as urban miming, phytomining, agromining and ect. Phytomining is an innovative approach to recover valuable metals from tailing pounds meanwhile making a step for remediation. So far in Bulgaria there are more than 980 Mt mining wastes. Based on the well-developed mining industry and temperate climate Bulgaria have a good potential for phytomining of noble metals.

5. References

[1] Sheoran, V., Sheoran, S., Poonia, P., 2013. Phytomining of gold: a review. J. *Geochem.Explor.* **128**, 42–50. https://doi.org/10.1016/j.gexplo.2013.01.008.
[2] Kostadinova, A, E Todorova, T Krumuv, Opportunities for stabilization of forest roads using waste from coal mining, *Physics and mathematics*, ISSN 2518-167X
[3] Todorova, E, A Kostadinova 2019. Risks of environmental pollution from mining waste from ore-containing copper processing, *Journal of environmental protection and ecology* 20(1):397-403
[4] Robinson, B., Anderson, C., 2021. In: van der Ent, A., Baker, A.J.M., Echevarria, G.,Simonnot, M.-O., Morel, J.L. (Eds.), Agromining of Thallium and Noble Metals BT - Agromining: Farming for Metals: Extracting Unconventional Resources Using Plants, *Springer International Publishing* 415–423 https://doi.org/10.1007/978-3-030-58904-2_20.
[5] Prabakaran, K., Li, J., Anandkumar, A., Leng, Z., Zou, C., Du, D., 2019, Managing environmental contamination through phytoremediation by invasive plants: *A review, Ecological Engineering,* **138**, 28-37
[6] National Strategy for Development of the Mining Industry, 2015
[7] BAMI, Metallurgy in Bulgaria, 2020
[8] Popov, P., Strashimirov, S., Popov, K., Kanazirski, M., Bogdanov, K., Radichev, R., Dimovski, S., Stoykov, S., 2012, Geology and metallogeny of the Panagyurishte ore region, *UMG „St. Ivan Rilski"*, ISBN 978-954-353-181-3
[9] Strashimirov, Str., Bogdanov, K., Popov, K., Kehayov, R. 2003. Porphyry Systems of the Panagyurishte Ore Region. *Society of Economic Geologists Guidebook,* **36**, 47-78
[10] Hikov, A., 2005, Geochemistry of stroncium and other trace elements in hydrothermally altered rocks at Petelovo deposit, Central Srednogorie, *Geochem. Mineral.* Petrol, **42**, 95-112
[11] Bogdanov, B., 1987, The Copper Deposits of Bulgaria, Tehnika
[12] Popov, P., Popov, K., 2022, Metallogeny of Bulgaria, *BMGK*, ISBN 978- 619-92104-0-6
[13] Investment proposals for "wet zone" with integrated microbial fuel elements in the area of Chelopech tailings pond and "Separation of a site for preliminary storage of non-hazardous ferrous metal waste on the territory of DPMCH EAD", 2017

[14] Tokmakchieva, M., 2008, The role of deep hypergenesis for forming complex mineralization, *Annual of the University of mining and geology "ST. IVAN RILSKI"*, **51**, Part I, Geology and Geophysics, 62-67
[15] Petrov, P., 2019, Approaches to the Reclamation of the Eastern Embankment, Ellatsite Mine, *Avangard Prima*, ISBN: ISBN 978-619-239-295-6
[16] Abramson, G., Stoyanov, S., Ermolaev, A., Kerbelova, V., Kyuchukova, M., 1988, Geochemical model of the Chiprovtsi-Martin ore zone, *Journal of the Bulgarian Geological Society*, **LII**, part3, 40-49
[17] Dimitrov, R. , 1988, Lead-zinc deposits in Bulgaria, *Technique*
[18] Sgorigrad – Stava. Identiche sciagure / Sgorigrad – Stava. Identical disaster / „Згориград – Става. Идентични бедствия", изд. Edizioni Arca (Лавис, Италия), 2009, ISBN 978-88-88203-52-2
[19] Mihailov, I., 2010, Burgas Copper Mines, *Adular*
[20] Boyadzhiev, S., Georgiev, V., Obretenov, N., 2012, Regional distribution of anomalous geochemical fields and root mineralization on the territory of the country, *Annual of Sofia University "St. Kliment Ohridski",* book 1 - geology, **103**, 149-226
[21] Kerestedjian, T., R. At anassova, I. K. Bonev, V. Hristov, L. Yurukova, V. Arnaudov, A. Hadjiev, I. Apostolova, T. Meshinev, 2000, Environmental impact of the mining activity - acomplex investigation in the Erma Reka valey, Madan ore district. - Geochem., *Mineral. and Petrol.*, **37**, 111-126
[22] Dobrev, S., Kolkovski, B. , 2014, Trace elements in main ore minerals of lead-zinc deposits from Madan ore field, *Rare and trace elements in Bulgaria: geological, technological and ecological aspects*, 94-101
[23] Vangelova, V., Neykova, E. , 2012, Mineral and geological composition of the sediments in the Laky tailings dam (central Rhodopes), environmental impact, *Annual of Sofia University "St. Kliment Ohridski ", Faculty of Geology and Geography*, book 1, **103**, 89-112
[24] Hadjiev, A., 2014, Phosphogypsum near Dimitrovgrad - resources for extracting rare and trace elements, *Rare and trace elements in Bulgaria: geological, technological and ecological aspects*, 140-147
[25] Todorov, T., 2019, Tailings and Mining Activities - Environmental Problems Related to Water Pollution, Thesis
[26] Zheleva, E., P. Bozhinova, P. Petrov, 2014, Analysis of pollution developments of lands and soils in the area of Dundee Precious Metals Chelopech "
[27] Wilson-Corral, V., Anderson, C., Rodriguez-Lopez, M., Arenas-Vargas, M., Lopez-Perez, J., 2011, Phytoextraction of gold and copper from mine tailings with Helianthus annuus L. and Kalanchoe serrata L. *Miner. Eng.* **24** (13), 1488–1494. https://doi.org/10.1016/j.mineng.2011.07.014.
[28] Wilson-Corral, V., Anderson, C.W.N., Rodriguez-Lopez, M., 2012. Gold phytomining. A review of the relevance of this technology to mineral extraction in the 21st century, *Environmental Management* **111**, 249-57, doi: 10.1016/j.jenvman.2012.07.037.
[29] Msuya, F.A., Brooks, R.R., Anderson, C.W.N., 2000. Chemically-induced uptake of gold by root crops: its significance for phytomining, *Gold Bull.* **33 (4),** 134–137. https://doi.org/10.1007/BF03215491
[30] Lamb, A.E., Anderson, C.W.N., Haverkamp, R.G., 2001b. The induced accumulation of gold in the plants Brassica juncea, Berkheya coddii and chicory, *In Chemistry in New Zealand*, **65** (Issue 2), 34–36.
[31] Anderson, C.W.N., Stewart, R.B., Moreno, F.N., Gardea-torresdey, J.L., Robinson, B.H., Meech, J.a., 2003, Gold phytomining. Novel developments in a plant-based mining system. In: Proceedings of the Gold, *Conference: New Industrial Applications of Gold*, **2**, 35–45
[32] Anderson, C., Moreno, F., Meech, J., 2005. A field demonstration of gold phytoextraction technology. *Miner. Eng.* **18 (4)**, 385–392. https://doi.org/10.1016/j.mineng.2004
[33] Handayanto, E., Muddarisna, N., Krisnayanti, B.D., 2014. Induced phytoextraction ofmercury and gold from cyanidation tailings of small-scale gold mining area of west Lombok, Indonesia. *Adv. Environ. Biol.* **8 (5)**, 1277–1284.

[34] Krisnayanti, B.D., Anderson, C.W.N., Sukartono, S., Afandi, Y., Suheri, H., Ekawanti, A., 2016. Phytomining for artisanal gold mine tailings management, *Minerals*, **6** **(3)**, 1–11. https://doi.org/10.3390/min6030084.

[35] Gonzalez-Valdez, E., Alarcon, A., Ferrera-Cerrato, R., Vega-Carrillo, H.R., Maldonado-Vega, M., Salas-Luevano, M. A., Argumedo-Delira, R., 2018. Induced accumulation of Au, Ag and Cu in Brassica napus grown in a mine tailing with the inoculation of Aspergillus Niger and the application of two chemical compounds, *Ecotoxicol. Environ. Saf.* **154**, 180–186.

[36] Szőllősi, R., 2020, Nuts and Seeds in Health and Disease Prevention (Second Edition), Pages 357-364, https://doi.org/10.1016/B978-0-12-818553-7.00025-5

[37] Tagami, K., Uchida, S., 2010. Rhenium: radionuclides. In*: Encyclopedia of Inorganic and Bioinorganic Chemistry*, 23–26

[38] Kabata-Pendias, A., 2011. Trace Elements in Soils and Plants, Fourth. CRC *Press, Boca Raton, Florida*. https://doi.org/10.1201/b10158-25.

[39] Hikov, A., 2014, Rare and scattered elements - distribution, classifications, application and more important types of localities, *Scientific seminar "Rare and scattered elements in Bulgaria: geological, technological and environmental aspects"*, 24-31

[40] Tzvetkova, C., Novo, L., Atanasova-Vladimirova, S., Vassilev. T., 2021, On the uptake of rhenium by plants: Accumulation and recovery from plant tissue, *Journal of Cleaner Production*, 1-8.

[41] Laubie, B., Vaughan, J., Simonnot, M.-O., 2021. In: van der Ent, A., Baker, A.J.M., Echevarria, G., Simonnot, M.-O., Morel, J.L. (Eds.), Processing of Hyperaccumulator Plants to Nickel Products BT - Agromining: Farming for Metals: Extracting Unconventional Resources Using Plants, *Springer International Publishing*, 47–61, https://doi.org/10.1007/978-3-030-58904-2_3.

[42] Li, Y.M., Chaney, R.L., Brewer, E.P., Angle, J.S., Nelkin, J.P., 2003. Phytoextraction of nickel and cobalt by hyperaccumulator Alyssum species grown on nickel-contaminated soils, *Environ. Sci. Technol.*, **37**, 1463–1468.

[43] Robinson, B., Brooks, R., Howes, A.W., Kirkman, J.H., Gregg, H., 1997, The potential of the high-biomass nickel hyperaccumulator Berkheya coddii for phytoremediation and phytomining, *Journal of Geochemical Exploration,* **60**, 2, 115-126

[44] Nemutandani, T., Dutertre, D., Chimuka, L., Cukrowska, E., Tutu, H., 2006. The potential of Berkheya coddii for phytoextraction of nickel, platinum, and palladium contaminated sites, *Toxicol. Environ. Chem.* **88** **(2)**, 175–185. https://doi.org/10.1080/02772240600585842.

[45] Kinska, K., Kowalska, J., 2019. Comparison of platinum , rhodium , and palladium bioaccumulation by Sinapis alba and their influence on phytochelatin synthesis in plant Tissues, *Pol. J. Environ. Stud.*, **28** **(3)**, 1735–1740. https://doi.org/10.15244/pjoes/89507.

[46] Rentsch, L., Aubel, I.A., Schreiter, N., Hock, M., Bertau, M., 2016. PhytoGer: extraction of germanium from biomass—an economic pre-feasibilty study. J *Bus. Chem.* **13**

[47] Tensho K, Yeh K-L (1972) Tracer study on preferential uptake of germa- nium by rice plant, *Soil Sci Plant Nutrition,* **18**, 173–178

[48] McMahon M, Regan F, Hughes H (2006) The determination of total germanium in real food samples including Chinese herbal remedies using graphite furnace atomic absorption spectroscopy, *Food Chem* **97**, 411–417

[49] Tyler G, Olsson T (2001b) Plant uptake of major and minor mineral elements as influenced by soil acidity and liming, *Plant Soil,* **230**, 307–321

[50] Reeves, R.D., Baker, A.J.M., Jaffre, T., Erskine, P.D., Echevarria, G., Ent, A., 2018.A global database for plants that hyperaccumulate metal and metalloid trace elements. *New Phytol.* **218** **(2)**, 407–411. https://doi.org/10.1111/nph.14907.

AUTHOR INDEX